IMAGES
of America

KING CITY

Smith and Hays. One of the first service stations in King City was Smith and Hays, located about midway down the south side of the 700 block of Broadway directly across from the high school. Pictured here on the left is owner George Hays. Notice the "Welcome to King City" sign along with the distances to Salinas, Gilroy, San Jose, and San Francisco. (Courtesy Carol Smith Sundahl.)

On the Cover: King City Auto Camp. This business, which included the Sunshine Garage and Service Station, was opened in May 1923 by L.C. Johnson. It was leased from F.E. Gause, who had recently built the facility at the northwest corner of First and Pearl Streets. In 1922, Gause sold his interest in the El Camino Garage to his partner W.J. Hamilton and concentrated on his auto camp. (Courtesy Michael J. Semas.)

IMAGES
of America

KING CITY

Howard P. Strohn, John R. Jernigan,
and Karen Vanderwall Jernigan

ARCADIA
PUBLISHING

ISBN 978-1-4671-0804-1

Published by Arcadia Publishing
Charleston, South Carolina

Printed in the United States of America

Library of Congress Control Number: 2022936899

For all general information, please contact Arcadia Publishing:
Telephone 843-853-2070
Fax 843-853-0044
E-mail sales@arcadiapublishing.com
For customer service and orders:
Toll-Free 1-888-313-2665

Visit us on the Internet at www.arcadiapublishing.com

This book is dedicated to Minerva "Minnie" Gervacio Strohn of Priest Valley for her ongoing support and interest in our effort to create this pictorial history of King City.

Contents

Acknowledgments

The authors are grateful for the cooperation and support from the Monterey County Agricultural and Rural Life Museum staff at San Lorenzo Park in King City; the San Antonio Valley Historical Association; the Monterey County Historical Society in Salinas; the Nacitone Museum in Lockwood; the King City Branch of the Monterey County Free Libraries; the City of King; the Celebrating Heritage branch of King City in Bloom; the King City and Southern Monterey County Chamber of Commerce & Agriculture; the *King City Rustler* newspaper; the Mayo Hayes O'Donnell Library, Monterey History and Art Association in Monterey; Martha Heasley Cox Center for Steinbeck Studies, San Jose State University; the National Steinbeck Center in Salinas, California; photograph collectors the late Pat Hathaway of Monterey, Michael J. Semas of Hanford, and Allen King of King City; photographer Alex Arce, former owner of King City Photo Center; and proofreaders Suzanne and Chuck Krause, Mauvoreen Nolte, Phoebe Cheney, John and Janet Buttgereit, Jim Albanese, Doreen Liberto, Maricruz Aguilar, Sharon Turner, Jessica Potts, Eron Daniels, Steve Wilson, Ken Erickson, Erica Sonne, Patricia Woodfill, Ann Beckett, Lee "Butch" LaVelle, David Sundahl, Susan Raycraft, Bob and Phyllis Walton, Donald Kohrs, Lisa Josephs, Peter Van Coutren, Jackie and Peter Anderson, Dayna Jacobs, Steve and Rose Adams, Mike LeBarre, Terry Lewis, John Pekema, Margaret Clovis, James Perry, Virginia Vanderwall, Melissa Jernigan, Greg Gallagher, Brandon and Kari Jernigan, Pat Robertson, Richard Casey, Anita Leonard, William "Bart" Bartosh, Nancy Jernigan Raymond, and J.T. Rethke.

Unless credited to others, all photographs in this book are from the collection of John and Karen Jernigan.

Introduction

Most people know King City for its location on California's historic Highway 101. However, the story of its beginnings leads to a deeper understanding of how this Salinas Valley town has evolved from a windswept desert into a rural community surrounded by agricultural fields responsible for helping feed the world.

Simply stand in downtown King City and gaze to the west to see the foothills and peaks of the Santa Lucia Mountains just over the ridge from the Pacific Ocean. Then look to the east and see the dry Gabilan Range of hills separating the Salinas Valley from the Diablo Mountain Range and the great Central Valley. Gaze north toward the Chalone Peaks, below which are the mostly flat terrain where the Salinas River flows north. The wind begins most afternoons, blowing south from Monterey Bay.

Just to the south of King City is San Lorenzo Creek, dry most of the year but raging at times during the winter as waters from the east merge with the Salinas River. San Lorenzo Creek was the southern border of the San Lorenzo (Soberanes) Land Grant in 1841 after Mexico divided the lands of Alta California into ranchos.

In this setting in 1886, capitalist Charles Henry King of Oakland took advantage of the future laying of the rail lines by buying acreage north of San Lorenzo Creek, knowing of the railroad's plan to extend south of Soledad.

From then on, with its unique location in the south of Monterey County, King City became a focal point for roads that led into surrounding areas like the hub of a wheel: areas called Jolon, Pine Canyon, Lockwood, Bitterwater, Peach Tree Valley, Priest Valley, San Lucas, San Ardo, and to the north, Greenfield. All of these originally brought trade to the town.

Over the years, King City became the launching point for the Los Padres National Forest to the west, which is generally covered with wildflowers in the spring. Fort Hunter Liggett, on land originally owned by William Randolph Hearst, is to the southwest along Jolon Road and surrounds Mission San Antonio de Padua. It is the third oldest mission in California, founded in 1771 by Father Junipero Serra, and is still preserved in its original setting. Farther down near Jolon Road are Nacimiento and San Antonio Lakes, vital to the storage of mountain water used by area farmers. Traveling north on Metz Road leads to the west entrance to Pinnacles National Park. A drive east on Bitterwater Road leads to the east entrance.

This book is an effort to celebrate the past with images that reflect the character of the residents who have created a rural community. Brought together by people of all cultures with hard work and a generous spirit, they made the town of 14,000 what it is today.

There were milestones along the way that greatly impacted the town. One of the first was the success of wheat crops planted by early landowners and Charles King. This proved that the "great Salinas Desert," as it was known on old maps, could be transformed into the patchwork of lush agricultural fields that fill the valley today. All this was thanks to irrigation.

Other crops greatly influenced the town over the years, including barley, sugar beets, pink beans, tomatoes, onions, garlic, carrots, lettuce, and spinach. The cattle and dairy industries also played important roles in shaping the town.

Much of what we know about the history of King City is thanks to the longevity of the *Rustler* newspaper, started in 1901 by Fred Vivian, who arrived in town on a bicycle and put together a printing press. Vivian saw the potential for the town, and for 44 years devoted himself to promoting its prosperity. He advocated for incorporation in 1911 and supported the installation of electricity, sewers, and a fire department. He also encouraged residents to shop locally to keep dollars in town.

One of the early merchants in King City was William J. Hamilton, the uncle of Nobel Prize–winning author John Steinbeck. Steinbeck's grandparents owned a ranch south of King City and his mother, Olive Hamilton, taught school locally until she met and married his father, John Ernst Steinbeck. The elder Steinbeck managed a grain warehouse for Southern Pacific Milling Company next to the railroad tracks. In his book *East of Eden*, Steinbeck makes 48 references to King City, most of which are based on his family history.

Another milestone in the life of the town was the creation of a US Army Air Force flight training detachment at Mesa del Rey Airport on Bitterwater Road during World War II. It is reported that 10,000 recruits trained here to determine if they would go on to be pilots during the war. The influx of servicemen greatly affected the community. Many eligible young men came to King City and married local women, staying the rest of their lives. A USO club transformed the town's nightlife until the war ended.

Highway 101 greatly impacted the town as a stopping point between Los Angeles and San Francisco. The route, also known as El Camino Real, ran directly down the main street, Broadway, after making a 90-degree turn off of First Street. From there it was gas stations, restaurants, and hotels that served travelers and brought dollars to the local economy. In the early days, vehicle accidents were a common occurrence that required the services of doctors and repair shops. Some travelers even took advantage of the numerous bars that provided liquid refreshments. The town was transformed in 1968 when Highway 101 bypassed Broadway but stayed east of the Salinas River, allowing businesses to relocate and still serve travelers.

Over the years, King City has grown to include a nine-hole golf course, the Mee Memorial Hospital, a swimming pool complex, a Hartnell College extension, the Salinas Valley Fair with the Rava and Topo event centers and rodeo grounds, a skateboard park, and soccer and baseball fields. One of the town's treasures is the Robert Stanton Theatre, built between 1937 and 1939 as a Works Progress Administration project. Architect Robert Stanton designed the massive concrete building that sits on the King City High School campus on North Mildred Avenue. Famed sculptor Jo Mora created the artwork on the front of the building. San Lorenzo Regional Park showcases the town's agricultural history in exhibits at the Monterey County Agricultural and Rural Life Museum.

King City is an hour from the larger shopping hubs of Paso Robles to the south and Salinas to the north, so residents have had to learn to live without some of the conveniences of bigger cities. In general, they have come to appreciate the less-hurried lifestyle, the fresh air, and the chance to get to know their neighbors.

It is our hope that readers will find more appreciation of their surroundings by better understanding King City's past through the stories and photographs presented here.

One

Rails Arrive, 1886

Charles H. King saw the potential for development in the Salinas Valley when the Southern Pacific Railroad started plans to lay tracks through what became the King Ranch on the San Lorenzo Land Grant. In 1884, King, an Oakland capitalist, purchased 13,000 acres for about $105,000 and planted a wheat crop that flourished. The construction of a rail line from Soledad south, mostly by Chinese laborers, arrived at the spot just north of San Lorenzo Creek in July 1886. The stopping point was labeled "Kings" or sometimes "King's Station." It was the start of Kings City, later shortened to King City.

Charles King was the man responsible for laying out the town in a one-square-mile plot. One story goes that King's workers burned off a section of stubble grazed by hogs to create space for the town, thus giving it an early nickname of "Hogtown."

It was Charles King's decision to invite his widower older brother Lewis to manage his King Ranch, which provided the city with some significant roots. Charles King's roots were in Oakland, and it appears neither he nor his family spent much time in King City, but Lewis King and his descendants did. Two of King's sisters, Mary King and Elizabeth King Bowden, made King City their home and are buried in the King City Cemetery.

One of the first buildings in King City was a grain warehouse that was built next to the railroad tracks. This was followed by businesses along Broadway and other dirt streets. The streets were choked by dust in the summer and muddy ruts during the winter. The daily railroad schedule played an important role in the delivery of goods to King City in the era before cars.

King continued to sell lots for residences and businesses for 13 years until he sold 11,906 acres of his ranch to Spreckels Sugar Company in 1897 for approximately $300,000.

More detailed information about the beginnings of King City can be found in the booklet published for the 100th anniversary of the town in 1986 and on the website of the Monterey County Historical Society. Articles in the *Salinas Index* newspaper also detail the arrival of the train in July 1886.

ENGINE 88 AT DEPOT. This could be one of the earliest photographs of the train depot in Kings City. Based on the style of the train, a historian from the California State Railroad Museum in Sacramento estimated the date to be 1889–1891, which was just three years after the founding of the town. Engine No. 88 was called the "Hurricane." Note the 29 people observing the action.

FOUNDER OF KING CITY. Charles H. King is shown here with his wife, Kate Brown King, in October 1900. Sixteen years earlier, he had purchased the San Lorenzo Rancho. The railroad was pushed south from Soledad and reached the site of King's proposed town in July 1886. In 1897, King sold his holdings to Spreckels Sugar Company. He died in Oakland on August 20, 1910, at the age of 68. (Courtesy City of King.)

RAILROAD EXCHANGE HOTEL. Located on the southwest corner of Broadway and First Street, this hotel was built in the late 1880s by F.W. Schroeder. By 1902, Peter Morasci became the owner, and by the fall of 1912, it was purchased by Emil Reich and became the Reich Hotel (pronounced "reach"). In 1928, it was moved to the northwest corner of First and Bassett Streets before being torn down in the early 1970s. (Courtesy Samuel G. Rice Jr. family.)

SOUTHERN PACIFIC GRAIN WAREHOUSE. Constructed in 1887, the year after King City was founded, this photograph shows this historic building on North First Street around the turn of the 20th century. Wheat and livestock were brought in from neighboring communities to be shipped. Eventually, it was sold and became the El Rey Milling Company. Later, it was Meyer Tomatoes. In 2017, the southern part was remodeled for H-2A farmworker housing. (Courtesy Allen King.)

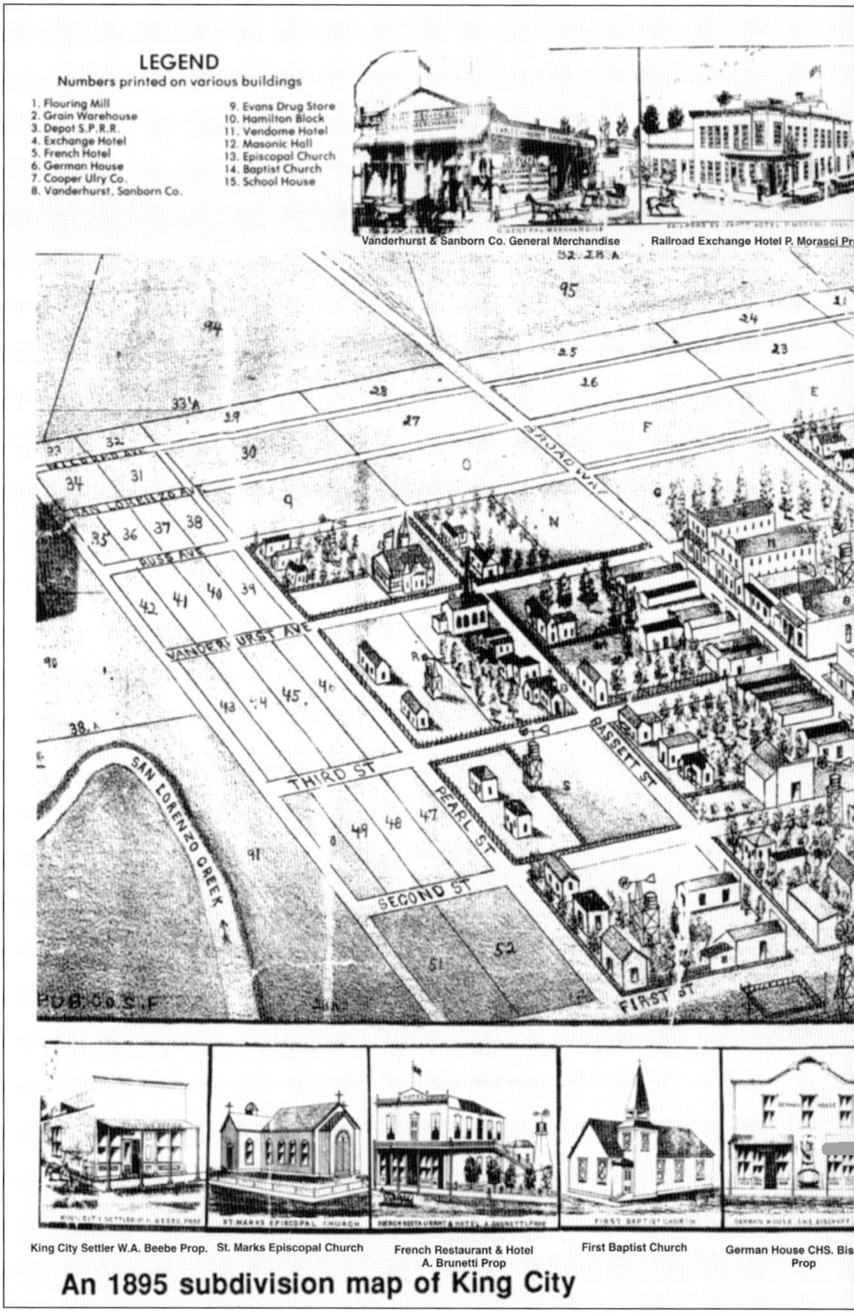

Burbank & Devendorf Map. This 1895 subdivision map shows the historic district of King City along with the 1,200 acres surrounding the original lots. Shown here with buildings are the 16 blocks laid out by William Minto in 1886. When city founder Charles King decided to sell his King Ranch, he hired Burbank and Devendorf to map out the areas north, west, and south of the town. The southern boundary was moved from Pearl Street to Division Street. The western city limits were moved from San Lorenzo Avenue to Canal Street. To the north, the boundary moved from Ellis Street to King Street. New lots were numbered, but Minto's map used letters. Interesting features of this map are the drawings of the significant buildings at the time as shown

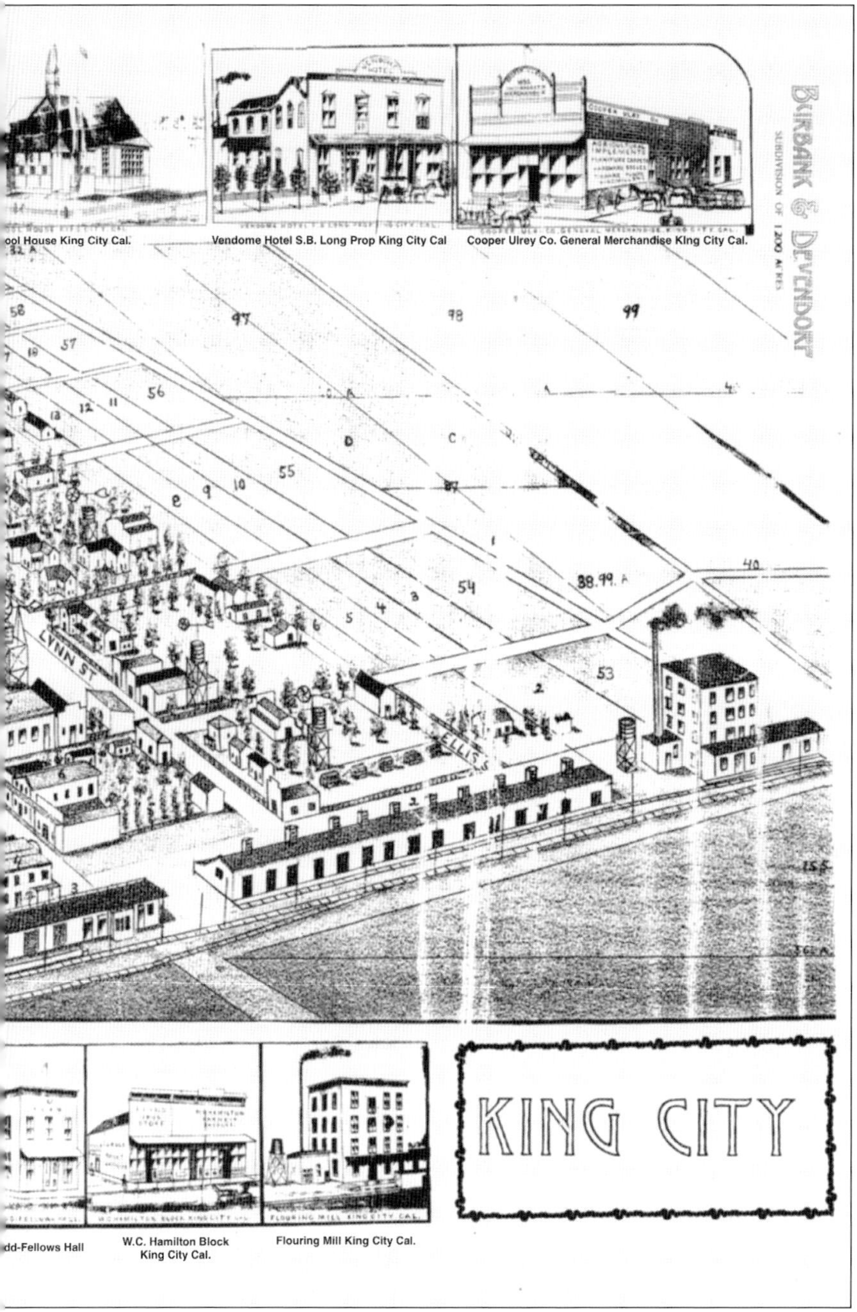

in the legend at top left. Buildings that still remain standing in 2022 are the Episcopal church (St. Mark's at 305 Bassett Street), the Baptist church (Assemblea Apostolica Church at 209 South Vanderhurst Avenue), the grain warehouse on the 100/200/300 blocks of North First Street (now seasonal employee housing/former Meyer Tomatoes building), the depot for Southern Pacific Railroad (moved to San Lorenzo Park) and Vanderhurst and Sanborn Company at 302/304/306 Broadway (now Pacific Valley Bank and La Vaquera Western Wear). Additionally, the building at 132 Broadway is one of the oldest structures in town. It was originally built in 1890 as Charles Conrad's Boots and Shoes. (Courtesy San Antonio Valley Historical Association.)

Depot and Hotel. This early photograph shows a scene looking southwest from the railroad tracks across First Street at Broadway with the depot on the left and the Railroad Exchange Hotel on the right. The hotel was founded by F.W. Schroeder in late 1886, the year before the depot was built. This photograph was taken by James E. Meddaugh of Watsonville, whose work appeared on early postcards of the area.

Kings City Depot. Here is a view from the north with a passenger train stopped for riders. "Kings City" is on the building. The picture is estimated to be before 1911, which is when the "s" was removed from the city name. The surrey with horses could well be Samuel Long, proprietor of the Vendome Hotel, waiting to transport guests. He was famous for his erratic style of driving. (Courtesy Michael J. Semas.)

Depot at King City. Residents gather to greet a passenger train in this view from the north. The parked vehicle indicates this photograph could be dated around 1915. The sign on the depot no longer shows Kings City but rather King City. Note the electric lights as the King City Water, Light and Power Company had "turned on the juice" in 1909. The girl's dress shows the famed "Salinas Valley wind" is up to its summertime habit. (Courtesy Bonnie Jean Grab.)

Moving the Depot. A piece of King City's history was preserved on June 2, 1989, when the abandoned depot building was transported down Broadway to a new home in San Lorenzo Park. Arranged by Robert L. Meyer of Meyer Tomatoes, the building was given to the care of the Monterey County Agricultural and Rural Life Museum, whose members used a $31,000 grant from the Harden Foundation to restore it for public use. (Courtesy Monterey County Agricultural and Rural Life Museum.)

Ford and Sanborn Store. This 25-by-100-foot structure at the northwest corner of Broadway and Third Street is shown as it appeared sometime before 1908. Founded in 1886 as Sanborn, Porter and Co., it became Vanderhurst and Sanborn and then Ford & Sanborn Co. It housed a hardware department and then a grocery store and a dry goods department. Note the windmill at the rear of the building.

Brunetti Store. Located on the south side of Broadway mid-block between First and Second Streets, this business was built in 1894 by Augustus Brunetti for his son Julius ("J.E."). Julius's brother Fred ("F.C.," also known as "Keeko") Brunetti returned to King City in 1911 to help with the grocery business that his brother started. Pictured here are Angelina "Angie" Brunetti and her brother Julius, with Jim Barton and Ernie Long.

A. WINCKLER GENERAL MERCHANDISE. This store was one of the first buildings in King City. Adolph Winckler came to King City in 1888 and set up a tent to conduct his business. He partnered with his son-in-law Joseph Rice, who was married to his daughter Sophie. In 1906, S.S. Hill became the owner of the Headquarters, a saloon inside the building. In 1936, it was replaced by the City Café building at 223 Broadway. (Courtesy Samuel G. Rice Jr. family.)

MEAT MARKET. Before Guadalajara Restaurant occupied 211 Broadway, this was the site of King City's first meat market. Charles Wilson bought lot 176 on February 23, 1887, and had this building erected. It was used as a meat market until 1926 with many proprietors, including Tom Treganza, James Lowe, Byron Norris, Bert Bray, Bert Rist, George Bryon, and Ray Benadom. In 1930, the building was torn down and replaced with a concrete structure.

SAINT MARK'S, KING CITY'S OLDEST CHURCH. City founder Charles King saved three lots on Bassett Street when he laid out plans for his new town. A testimony to the importance of spiritual values in the lives of residents, the Episcopal, Baptist, and Catholic denominations were the recipients of these parcels. The Episcopal church at 305 Bassett Street held its first service here on Easter 1890 under the leadership of Rev. James Shannon McGowan.

CATHOLIC CHURCH ON BASSETT. Built in 1891 on the northeast corner of South Russ Avenue and Bassett Street, the Catholic church was given the name St. John the Baptist. For more than 50 years, it was the scene of many social gatherings including baptisms, confirmations, weddings, and funerals for people of all nationalities, including many Swiss-Italian and Mexican families.

Baptist Church, 1926 Funeral. This is King City's third oldest church, at 209 South Vanderhurst Avenue at the corner of Bassett Street. Although the steeple has been removed, residents have known this building as Community Baptist Church and Community Church. As of 2022, it offered Spanish-language services to the Hispanic community as Assemblea Apostolica. This photograph was taken in 1926 during the funeral of police officer Clarence Dodge, who died in a motorcycle accident on Broadway.

Moving the Catholic Church, September 1948. These buildings were moved from Bassett Street to become part of the church's current and larger campus at 504 North Third Street. This photograph shows how it was done in those days: put the buildings on wheels and pull them down Broadway. Note the workers riding atop the building at the corner of Vanderhurst Avenue and Broadway heading north. (Courtesy Dan and Celie Hayes family.)

J.C. Annear Blacksmithing. This business was started by John Meeter in the late 1890s near the northwest corner of South Second and Bassett Streets. In 1903, it was sold to the Dayton brothers. J.C. Annear purchased the interest of his partner W.R. Dayton in 1908. Eventually, Annear ran a saloon on the southeast corner of Broadway and Second Street. (Courtesy Michael J. Semas.)

Svart Blacksmith Shop. Nis Andersen Svart established his blacksmith shop on Broadway on a lot purchased from Charles H. King in August 1887. Located on the south side in the middle of the 300 block, the business was next to the Gem Saloon. A native of Denmark, Svart arrived in the United States in 1875. He married Anne Hansen and lived at 316 Bassett Street. This image was provided by his great-granddaughter. (Courtesy Sandy Holmes Kattelman.)

King City Livery Stable. Tom Pettitt acquired this property west of Hotel El Camino in July 1908. By September, the foundation and floor were finished, and he hosted a dance. Two months later, William Koester signed a long-term lease. In October 1910, S.C. Corey and R.R. Webb took over. William J. Hamilton and Frank Gause remodeled the building in September 1915 to house their Ford dealership, at 116 Broadway.

Blackhawk Stables. Built for the Phillips brothers in the summer of 1901, this building was sold to R.R. Diaz of Jolon in 1905, who put Johnny Leonard in charge. Once livery stables were no longer needed, the building had many uses, including as the Hayloft during World War II, a place of entertainment for soldiers. In 1926, the *Rustler* reported on "Blackhawk Stables, whose disappearance symbolizes the downfall of the horse in King City." (Courtesy Dan and Celie Hayes family.)

First Vendome Hotel. Originally a wooden structure at 320 Broadway, this hotel was built by Vanderhurst, Sanborn Co. in 1886 and was originally known as San Lorenzo Hotel. It appears that A.L. Tynan of Salinas was the first proprietor. King City's first election in November 1886 was held here with C.H. King as the inspector and James Lynn and N.A. Svart as judges. Other early owners were J.S. Cronk, S.E. Moore, Leon Ticknor, McKinsey and Rutherford, and Emil Reich.

Vendome Hotel Fire. July 17, 1914, was a dramatic day in the life of King City, with the fire that destroyed this downtown building. Intense heat cracked the plate-glass windows on the Ulrey Garage and the *Rustler* across the street on the 300 block of Broadway. All adjoining structures were saved by diligent firefighters. The owner and proprietor, Gustav Bahm, erected a temporary saloon to generate income while he planned to rebuild the hotel.

Reinforced Concrete Vendome Hotel. By 1918, ground was broken for the rebuilding of the hotel, and six months later, an announcement appeared in the *Rustler* stating the new hotel was open to guests. It was described as a "splendid concrete structure." By 1921, owner Bahm leased rooms over the *Rustler* office across Broadway to keep up with demand. The hotel is remembered as the location of the famed Top Hat Cocktail Lounge. (Courtesy Robert "Bobby" Mansfield.)

Vanderhurst Hall. Merchant William Vanderhurst had this building constructed in 1889 at the northeast corner of Broadway and Vanderhurst Avenue. Early on, it hosted activities of the San Lorenzo Social Club with the lodge of Free and Accepted Masons in the upper story. The trees were referred to as "the park" in early newspaper accounts. As of 2022, the site is slated for a future downtown plaza. (Courtesy Michael J. Semas.)

ODD FELLOWS HALL, KINGS CITY. Photographer R.W. Horne captured this scene of the south side of the 100 block of Broadway sometime around 1905. The IOOF Hall was also called "the Opera House." Entertainment was provided downstairs, and lodge meetings were upstairs. The building to the left was the Opera House Bar, started by McKinsey and Rutherford. In 1909, Norris Logan remodeled the building and advertised the "swellest" gin dispensary in town. (Courtesy Monterey County Agricultural and Rural Life Museum.)

INSIDE IOOF HALL. This formal gathering at the hall of the International Order of Odd Fellows is significant for the business signs shown on the back wall. Based on them, it appears this photograph was taken in 1908 or 1909. Names include J.N. Besse & Co. Real Estate, Blackhawk Livery Stables, Monterey County Savings Bank, J.C. Annear Blacksmithing, and S.S. Hill Headquarters saloon. (Courtesy Monterey County Agricultural and Rural Life Museum.)

CHIEF SALOON. Henry Anderson's business at the northeast corner of Broadway and Second Street was a one-story structure built in 1890 for Charles Conrad's Boots and Shoe Store. On the right side of the building was Gus Burchard's barbershop. Anderson opened the saloon in 1903 and, three years later, bought the property from Conrad. The city council declined to renew Anderson's bar license in 1921, thus closing his business.

THE AUTO. Saloons like the one featured in this ad played a significant role in early King City. Charles Howell operated the Auto at 315 Broadway. When the west side of the present building became King City's first garage, Howell was operating his "resort" on the east side of the building under the name of Café Royal. Once the garage opened, he decided naming it "the Auto" was more appropriate.

CEMETERY MAP, 1898. The Kings City Cemetery Association filed its articles of incorporation in January 1898 and a subdivision map was drawn up in July. In 1899, the association sold the 7.543 acres to San Bernabe Lodge No. 358 of the International Order of Odd Fellows, which ran it for 40 years. In 1939, the King City Cemetery District was formed, and directors purchased the property from the IOOF. The district continues to oversee the cemetery.

HORSES AND BUGGIES AT CEMETERY. This burial at the King City Cemetery at 1010 Broadway could be around 1906. Note the telephone poles and lines that were activated to Jolon in 1905. Early on, the west side of the cemetery was for Catholics and the east side for Protestants. The Salinas River is flowing full, which indicates it was winter. (Courtesy Monterey County Agricultural and Rural Life Museum.)

Two

What Came Before

The Native American Salinan people inhabited the lands around what is today King City before 1769. They left us with petrographs in the caves to the west. Little else survived except the stories passed on by their descendants, families with names such as Encinales, Laguna, and Pastor.

In 1769, Spanish explorer Gaspar Portola and his expedition passed near the site of present-day King City looking for Monterey Bay to establish a presidio. Catholic father Juan Crespi was with Portola, and Crespi's diary gives information about the trek to meet Father Junipero Serra in Monterey. Father Serra traveled south from Monterey in 1771 to establish Mission San Antonio de Padua. During this era, Mission Soledad and Mission San Miguel were also established, putting King City right on the route sometimes referred to as El Camino Real—the King's Highway.

In 1776, Juan Bautista DeAnza and his expedition of 240 men, women, and children passed through the area while establishing a land route through Alta California.

In 1822, Mexico achieved independence from Spain and assumed jurisdiction over California. The land was then divided into ranchos. In 1841, Feliciano Soberanes was granted ownership of Rancho San Lorenzo. (Soberanes descendants are buried in the King City Cemetery.)

Records indicate that Soberanes sold Rancho San Lorenzo to Eugene Sherwood of Salinas in 1856, but drought and floods discouraged Sherwood. He sold the rancho in 1865 to Carlisle S. Abbott and David Jacks, who kept it until financial troubles forced them to sell. In 1880, records show the property was owned by R.M. Shackelford, Solomon Boswell, and Loren Coburn. Charles King was finally able to start buying Rancho San Lorenzo on October 13, 1884, when the Estate of Boswell agreed to sell to him.

Painted Cave. Evidence of the Salinan people's influence on the King City area remains in these paintings to the west on the property over the hill from Pine Canyon on Fort Hunter Liggett. Highly treasured by tribal descendants and historians, these markings are some of the only things left to document the era before photography. More information about Painted Cave is found in Beatrice "Tid" Casey's book *Padres and People*. (Courtesy San Antonio Valley Historical Association.)

The Encinales Family. Perfecta Encinales (first row, far left) was the matriarch of this Salinan family. Known as an exquisite basket maker, her husband, Eusebio, was equally well-known as a vaquero, or cowboy. Those with Salinan ancestry who live in the King City area can often trace their heritage back to her. She and her husband were devoted to San Antonio Mission de Padua in Jolon. (Courtesy Mayo Hayes O'Donnell Library, Monterey History and Art Association.)

King City Bean and Grain. Salinan Native American Dolores Encinales stands in front of this grain office with his trademark hat in hand. Located on the southwest corner of Broadway and Second Street, the grain business office was connected to William J. Redding, who served as Third District Monterey County supervisor from 1939 to 1959. Encinales is buried in the King City Cemetery. (Courtesy Mayo Hayes O'Donnell Library, Monterey History and Art Association.)

George and Ben Burch, Salinan Descendants. Native Americans like George Burch (right) and his brother Ben worked in King City businesses and served in the military. Their maternal grandmother, Maria "Mary" Govers, was a full-blooded Salinan who married Frank Blas McCormack. One of George's first jobs in 1938 was at a King City bakery. He then joined the US Marines during World War II. About the time George completed his military service, his brother Ben enlisted in the US Navy and served in Korea. Both are buried in the King City Cemetery. This image was provided by George Burch's daughter. (Courtesy Patti Burch Dunton.)

In the Shadow of Pinyon Peak. The tallest visible mountain to the west of King City is Pinyon Peak in Los Padres National Forest, most easily distinguished when covered in snow. Just beyond it to the west is the tallest point in the Santa Lucia Mountain Range, Serra Peak (elevation 5,873 feet), originally called Santa Lucia Peak. The name was changed in 1906 to honor Father Serra.

Salinas River near King City. The landscape that formed the Salinas Valley existed long before the town of King City, and the river played an important role in its settlement. Water is, and has always been, the lifeblood of the area, even though the river only runs aboveground part of the year. With headwaters near Santa Margarita, the Salinas River is joined by the San Antonio and Nacimiento Rivers south of Bradley. At King City, the waters of San Lorenzo Creek from the east and Pine Canyon Creek from the west join on a journey to the Pacific Ocean.

The Chalone Peaks. Guarding King City from the north are two mountains that are part of the Gablian Range. Situated east of Soledad, the Chalone Peaks tower over Pinnacles National Park and are visible from miles away. In 2007, King City officials named Chalone Peaks Middle School on the north end of town after the mountains.

Pinnacles National Monument. Although not visible from the city limits, this spectacular geological formation intrigued early explorers and has been promoted as a tourist attraction for those visiting King City. The area was made a national monument in 1908 but was redesignated as a national park by Congress in 2012. It serves as a place for hiking and climbing as well as being a home to the endangered California condor. This postcard is dated July 3, 1933.

Mission San Antonio de Padua. Founded by Father Junipero Serra on July 14, 1771, this mission is 25 miles southwest of King City. Construction on the building pictured here began in 1810 and was completed three years later. The mission was abandoned from 1882 to 1928. Various attempts at restoration were done; however, it was the restoration started in 1948 that produced an accurate replica of the 1813 original.

Dutton Hotel, Jolon. Located near Mission San Antonio 19 miles southwest of King City, this original adobe structure was built by Antonio Ramirez perhaps as early as the 1850s. In 1876, Thomas Tidball and George Dutton became partners and bought the Dutton Hotel, but a disagreement led to Tidball moving down the road. Dutton heirs sold the property to William Randolph Hearst in 1929.

Lowe's Stage Station. A native of Pennsylvania, James Lowe settled on property along Jolon Road about nine miles south of King City in December 1865. Eventually, his settlement included a store, the San Antonio Post Office, a flour mill, a barn, and stables for the stage teams. Lowe sold his property to C.H. Mansfield, and the land was later owned by the Gamboa and Bengard families. (Courtesy San Antonio Valley Historical Association.)

Site of Early Stage Stop. This is a 1946 photograph of the Bombay Club, a popular night spot on Jolon Road six miles southwest of King City. Originally it was known as Walker's Station. It was built for Andrew E. Walker, who homesteaded the property and settled there in September 1871. He conducted a hotel and store with a stage stop. (Courtesy Mayo Hayes O'Donnell Library, Monterey History and Art Association.)

Selleck Home, January 1878. This Bitterwater homesteader's cabin is typical of those that dotted the canyons and valleys around King City as settlers started arriving as early as the late 1860s. The priority for homesteaders was to have a permanent shelter to protect their families. In their homestead application, the house would be described by how many rooms, size of the rooms, number of windows and doors, and the presence of stairs. (Courtesy Selleck family.)

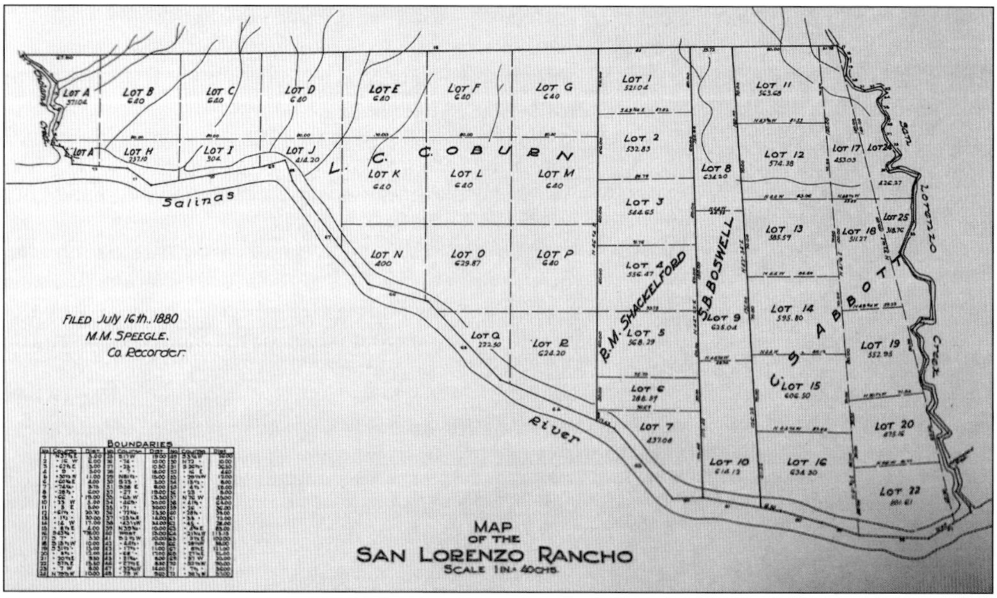

San Lorenzo Rancho Map. Filed July 16, 1880, in the Monterey County Recorder's Office, this map shows the southern boundary at San Lorenzo Creek and the northern boundary at Chalone Creek. The western border was the Salinas River, and the hills to the east roughly marked the eastern boundary. It shows the three parcels owned by R.M. Shackelford, S.B. Boswell, and C.S. Abbott that were purchased by C.H. King in 1884.

Three

A River Runs through It

The Salinas River provides a western border for King City. Since its founding, the river's influence has shaped the life of the town.

Known as "the upside-down river" in Anne B. Fisher's *Rivers of America*, the Salinas is famous both for flooding and being as dry as a desert, and for having a vast store of underground water. Water flowing out of the Santa Lucia Mountains to the west pours down the San Antonio and Nacimiento Rivers to meet the Salinas River just south of Bradley. Then it winds its way north to empty into the Pacific Ocean near Moss Landing.

The first bridges across the Salinas River at King City were regularly subjected to washouts. Jetties were installed to try to prevent erosion of the banks. For a while, a "flying duck" contraption featuring a basket on a cable was used to ferry people from the Pine Canyon area to King City when the bridge was washed out.

Eventually, the idea of damming the tributaries of the Salinas River took hold with the formation of Nacimiento Lake in 1956 and San Antonio Lake in 1965. This helped lessen flood damage and saved water for farming, but the river still often dried up and went underground, leaving the riverbed dry. Some old-timers used to bet on how long it would take for the Salinas to reach the Soledad Bridge once it flowed through King City in the winter.

In the first few pages of his book *East of Eden*, John Steinbeck describes the Salinas River this way: "It was not a fine river at all, but it was the only one we had and so we boasted about it—how dangerous it was in a wet winter and how dry it was in a dry summer. You can boast about anything if it's all you have. Maybe the less you have, the more you are required to boast."

Cattle Crossing River. A local rancher is fording the Salinas River with his herd of cattle. This was a common practice to move livestock from pasture on one side to the other. Quicksand was a danger, and at times, cattle and horses would sink to their bellies, making it hard to cross. There are reports of animals and wagons disappearing into the quicksand. (Courtesy Michael J. Semas.)

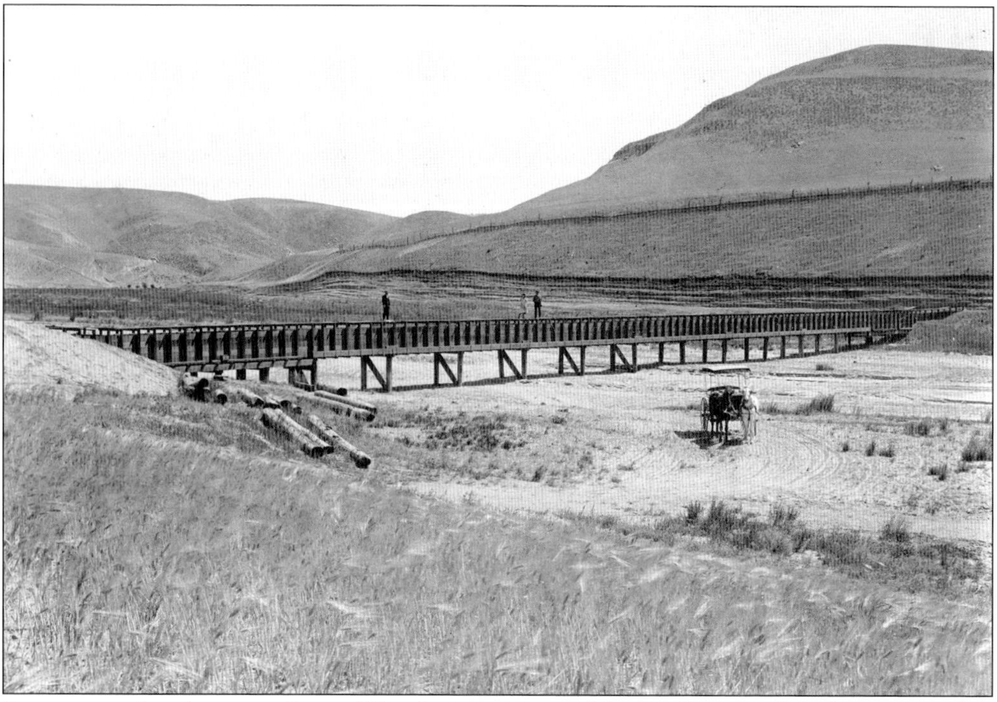

Flume over San Lorenzo Creek. This flume was part of the San Lorenzo Canal constructed in 1896 to divert water for irrigation. The canal was 8.5 miles. It began near Bitterwater Road where the valley reaches the foothills about five miles east of King City. Canal Street got its name from another canal, the Salinas Canal, which fed from the Salinas River near Wildhorse Canyon. (Courtesy Monterey County Agricultural and Rural Life Museum.)

SALINAS RIVER. Here is a view captured by Harlan of the King City Bridge being used by horses. It shows the downstream side with its rebuilt span on the left. This was the only road connection from King City for access to Pine Canyon, Jolon, and Three-Mile-Flat (an early name for Greenfield). A note on the back is dated August 22, 1907, and mentions "Mary's little lamb." Where did the sheep in the foreground come from? (Courtesy Michael J. Semas.)

KING CITY BRIDGE. This is a photograph by James E. Meddaugh of the east end of the King City bridge looking west across the Salinas River. The wooden planks that lined the floor of the structure are clearly visible. Also note the sign at the top that reads, "$10 to $100 fine for riding over this bridge faster than a walk." Postmarked April 1908, a message on the back states, "They say this bridge is 1/2 mile long."

FLYING DUCK. A boxed wooden car was used to cross the 500-foot gap, washed out by the February 1909 flood. The car was built by Joe Wasson, a local blacksmith, and was supported by a cable attached to the damaged end of the bridge. The other end was tied to a tree on the bank. In this photograph, it is being used to carry the mail to residents west of the river.

FLYING DUCK NO. 2. This device was used again in March 1911 after another flood created a 200-foot gap in the Salinas River Bridge. Writing on the image indicates the mail is being carried. Note the horse and wagon waiting on the bridge to pick up the mail for delivery. The importance of the bridge is shown by the power poles attached to get utility service west of King City. (Courtesy Howard P. Strohn.)

KING CITY BRIDGE. This steel truss bridge was completed in February 1910. In the foreground, the jetties to keep the Salinas River from cutting away the bank can be seen. Monterey County surveyor Lou Hare drew up the plans. John McReynolds was the contractor. This view is from the south. J.E. Riley, who had a photography studio in King City, captured this image. (Courtesy Monterey County Agricultural and Rural Life Museum.)

WORKERS ON BRIDGE. Looking from the northwest with King City in the background in 1910, workers are completing the last span of the Salinas River Bridge. The piers consisted of two five-foot-diameter steel cylinders driven 10 feet into the riverbed. Four 40-foot-long piles were driven into the riverbed, and the cylinders were filled with reinforced concrete. (Courtesy Monterey County Agricultural and Rural Life Museum.)

Broken San Lorenzo Bridge. Floodwater in San Lorenzo Creek in March 1911 caused this collapse on First Street. On the right is the railroad trestle and in the background is King City. An interesting sidelight is that local mechanic E.E. Mansfield drove across the unstable structure to San Lucas and returned with it collapsing behind him. The *Rustler* headline read, "Dare-Devil Mansfield and his Reo" above a photograph of him crossing. (Courtesy Howard P. Strohn.)

Wash Out near Metz. The Southern Pacific Railroad Bridge near Metz Road, 11 miles north of King City, washed out at the Chalone Creek crossing during the flood of March 1911. As the image indicates, rail traffic was halted for a time. Chalone Creek drains into the Salinas River from the land where Pinnacles National Park sits to the east. King City photographer C. Harlan took this photograph for a postcard.

Wagon on Bridge. Here is a c. 1911 view from the downstream side showing the rebuilt span. A note on the back points out that "the last two years the river has washed around back of the bridge and left it so no one could get across." A horse and wagon with two aboard can be seen crossing from King City to the west side.

Temporary Stores during Flooding. When floods washed out the east end of the bridge, businesses in King City would accommodate residents on the west side with a wagon like the one seen here. It held merchandise for people to purchase without crossing the river. The two main businesses that did this were A. Widemann and the H-A-F Co. and their predecessors Ford and Sanborn Co.

GOLF COURSE FLOODING. On multiple occasions, floodwaters from San Lorenzo Creek have inundated the King City Golf Course since it was built by the city in 1953. In this photograph, residents, including Jim Albanese (third from left), use shovels to fill sandbags in an effort to contain the rising waters. The nine-hole course is at the corner of Villa Drive and South Vanderhurst Avenue. (Courtesy King City Golf Course.)

FLOODING FROM SAN LORENZO CREEK. Rainfall in January 1969 surged into this wrecking yard on Jayne Street just east of the First Street Bridge near where it meets the Salinas River. There is no dam on San Lorenzo Creek, so runoff from the Peach Tree and Bitterwater Valleys has periodically damaged areas near the King City Golf Course. In this photograph, the owners of King City Glass & Body Works Inc. survey the damage.

Four

The John Steinbeck Connection

When a Nobel Prize–winning author bases one of his most famous novels on your town, people sit up and take notice. Set in the Salinas Valley, John Steinbeck's 600-plus page novel published in 1952, *East of Eden*, refers to King City 48 times. His descriptions of the area ring true to anyone who knows the Gabilan and Santa Lucia Mountains and small towns that dot the Salinas Valley.

Steinbeck based *East of Eden* on situations similar to those of his family. King City can claim Steinbeck's roots. His mother's family were the namesakes of Hamilton Canyon eight miles southeast of King City near Wildhorse Canyon. His grandparents Samuel and Elizabeth Hamilton came from Ireland and eventually established a homestead. Steinbeck's mother, Olive, was one of nine children and the youngest of five daughters. She became a school teacher, spending the 1889–1890 school year in King City.

Not far from the grammar school was the Southern Pacific Milling Company warehouse where Steinbeck's father, J. Ernst Steinbeck, worked as the first agent after the railroad arrived in town in 1886. Olive and Ernst were married in San Francisco on December 22, 1890.

Steinbeck was born on February 27, 1902, in Salinas after his parents left King City. However, the younger Steinbeck had many memories of visiting the Hamilton Ranch homestead. His uncle William J. Hamilton was a businessman. His aunt Dessie was a dressmaker who saw clients at the Railroad Exchange Hotel. His uncle Thomas was well-known in town.

Over the years, there has been confusion between Steinbeck's uncle William J. Hamilton and William C. "Will" Hamilton, who were prominent in town for many years. Both men were members of the Masonic lodge and the chamber of commerce. They were contributors to St. Mark's Episcopal Church and involved in civic affairs. William C. was a harness maker who came from Hollister, married Birdie King (niece of city founder Charles King), and served as city trustee and justice of the peace. William J. owned the El Camino Garage and H-A-F Company. His wife, Adela Winckler, died following the Spanish flu, and William J. went on to marry Lucy McCandless.

Steinbeck's uncle William J. Hamilton is buried in the Garden of Memories in Salinas near the Steinbeck family. William C. Hamilton and his family are buried in the King City Cemetery near the grave of Lewis King, Charles King's brother.

There is no relation between the two men, but because of their similar names, they have often been confused.

John Ernst Steinbeck. The father of author John Steinbeck was the son of John Adolph and Almira Steinbeck of Hollister. One of the first settlers in King City, he was appointed manager of the Central Milling Company flour mill (later the Southern Pacific Milling Company) at King City in February 1887 next to the newly arrived railroad. In September 1892, he transferred to Paso Robles and later to Salinas. In 1923, he was appointed treasurer of Monterey County. (Courtesy Martha Heasley Cox Center for Steinbeck Studies, San Jose State University.)

S.P. Milling Co. John Ernst Steinbeck worked at the Southern Pacific Milling Company as its agent and bookkeeper on North First Street. He claimed in a speech to the King City Rotary Club in August 1929, "I was the pioneer, the first man to discover the San Lorenzo Rancho as the ideal site for a town." He also worked for the company in Paso Robles and Salinas. (Courtesy Monterey County Agricultural and Rural Life Museum.)

Olive Hamilton. This 1890 formal photograph of John Steinbeck's mother was taken near the time she was working as a teacher in King City in the fall term of 1889. After passing the county board exam at age 17, she taught at Bernabe School southwest of King City, the Malpaso School south of Carmel, and the Fairview School District near Soledad. (Courtesy Martha Heasley Cox Center for Steinbeck Studies, San Jose State University.)

Research. Author John Steinbeck is seen here in February 1944 researching his upcoming novel *East of Eden*. He is reading the *Salinas Index*, which began publishing in 1873 and later became the *Salinas Californian* in 1942. *East of Eden* was set at his grandfather's ranch southeast of King City in Hamilton Canyon. Steinbeck grew up visiting family at the ranch. (Courtesy Martha Heasley Cox Center for Steinbeck Studies, San Jose State University.)

Miss Dessie Hamilton,

FASHIONABLE

DRESSMAKER,

SALINAS,

Will be at Schroeder's Hotel.

King City.

SATURDAY, NOV. 16,

And every two weeks thereafter, and will be prepared to take orders for any desiring dresses.

AUNT DESSIE, DRESSMAKER. This ad for Dessie Hamilton appeared on November 21, 1889, in the *Salinas Valley Settler* newspaper of King City. Dessie was a sister to Steinbeck's mother, Olive Hamilton, and reportedly lived with the Steinbeck family in Salinas. Schroeder's Hotel was also known as the Railroad Exchange Hotel, a mainstay for traveling salespersons. In March 1907, while staying at the Hamilton ranch southeast of King City, she fell ill and passed away at age 48.

EL CAMINO GARAGE. Construction began on this building on the north side of Broadway between First and Second Streets in fall 1917. It replaced the frame structure known as the Tom Pettitt King City Stables. Partners were Frank Gause of Jolon and W.J. Hamilton, uncle of John Steinbeck. In 1922, Hamilton purchased the interest of Gause and became the sole owner. (Courtesy Howard P. Strohn.)

Hamilton Family 50th Anniversary, 1899. Shown here is the family of Elizabeth and Samuel Hamilton gathered at the home of John and Olive Steinbeck in Paso Robles. Immigrants from Ireland, Samuel and Elizabeth lived in San Jose. They moved to the newly settled town of Salinas in 1871 before moving in August 1887 to their homesteaded ranch southeast of King City. From left to right are (children) Elizabeth "Beth" Steinbeck (daughter of Ernst and Olive), Marian Hamilton (daughter of Mamie and George), and Esther Steinbeck (daughter of Ernst and Olive); (first row) Olive Hamilton Steinbeck (John Steinbeck's mother), George Washington Hamilton, Adela Winckler Hamilton (W.J.'s wife), Elizabeth Fagen Hamilton, and Samuel Hamilton; (second row) John Ernst Steinbeck (John Steinbeck's father), Mamie Hamilton (wife of George), William John "W.J." Hamilton (husband of Adela), Joseph Hamilton, Lisette "Lizzie" Hamilton Farrel (wife of J.P.), Thomas Scott Hamilton, Adelia "Dessie" Hamilton, Will Martin (husband of Molly), and Molly Hamilton Martin. (Courtesy Martha Heasley Cox Center for Steinbeck Studies, San Jose State University.)

Hamilton Harness. The name Hamilton played a large part in early King City, and confusion between two different Hamilton families occurred. W.J. Hamilton, an uncle of John Steinbeck, was involved with the H-A-F Company and the Ford dealership. W.C. Hamilton, whose harness shop was on the southwest corner of Third Street and Broadway, was a native of Iowa and came from Hollister. He married Birdie King, niece of the town's founder. (Courtesy Monterey County Historical Association.)

Broadway Looking East, c. 1918. Anchoring the 300 block at the northwest corner of Third Street, the H-A-F Company was founded by W.J. Hamilton, Robert R. Allen, and Chris Falkenberg in 1910. W.J. was the uncle of John Steinbeck (not to be confused with W.C. Hamilton, who owned the Hamilton Harness Shop across the street). Hamilton, Allen, and Falkenberg were at one time employees of Ford & Sanborn Co., which later acquired the business.

Five

Turn of the Century

Perhaps the most significant influence on King City after 1900 was the arrival of a young newspaperman named Fred Vivian. Riding his bicycle from San Jose, he came to town with visions of using publishing to support his family and promote the fledgling town.

Records in the *Rustler*, starting with the first edition on May 11, 1901, have documented the town for 120 years. Vivian advocated for irrigation, paved streets, good schools, tourism, and all things for the betterment of the town. His newspaper documents the building boom of 1917, the impact of World War I, the Spanish flu epidemic, and the development of agriculture and irrigation.

Because his newspaper survived on advertising, he was staunchly pro-business and promoted shopping locally. His philosophy was reflected in a quote attributed to San Lucas pioneer Julius Trescony: "Buy at home, sell at home, stay at home, and you will always have a home."

The home Fred Vivian made for himself and his descendants has had a far-reaching impact on the fifth generation. Daughters Ruth and Beatrice were among the first to graduate from King City High School. Ruth went on to edit the *Rustler* with her husband, Bill Steglich, and was a champion for the beautification of the King City Cemetery. Daughter Beatrice "Tid" Casey became the author of *Padres and People*, the go-to book for the history of the Mission San Antonio de Padua. Grandson Harry Casey went on to run the newspaper and, like Fred Vivian, was a champion of all worthy local causes. He started Casey Printing, now operated by Casey descendants, and also wrote a trilogy of historical fiction novels centered on King City and its surrounding areas.

THE *RUSTLER* AT 323 BROADWAY. The third home of the *Rustler* was this building owned by Paul Talbott constructed by Charles Steiner. Founder and publisher Fred G. Vivian (shown at right) moved here in August 1908. In November, another structure on the left was incorporated into the building, nearly identical to this one. As of 2022, the buildings still stand. The *Rustler* moved to 110 North Vanderhurst Avenue in 1931. (Courtesy Casey family.)

RAILROAD CORRALS. This photograph is proof that sheep regularly occupied the northeast corner of First and Pearl Streets. Note the Reich Hotel and First Street in the background. Standing fourth from left is John Lacrabere, a well-known rancher and property owner in King City at the time. It is believed this photograph was taken around 1913 before the corrals were moved farther south in 1916. This image was provided by the grandson of John Lacrabere. (Courtesy Butch LaVelle.)

Eatondale Repair. In this April 1909 postcard, the building at right was the first automotive garage in King City, built by Paul Talbott. E.E. Mansfield bought the former Svart Blacksmith shop two years earlier and moved his business here to Broadway just east of Vanderhurst Avenue. A well-known local mechanic, Mansfield invented a self-aligning piston, which was manufactured in King City, and a new type of carburetor. Vanderhurst Hall is shown on the left.

Hotel Celebration. It is likely that this was an event to celebrate the grand opening of the annex of the El Camino Real Hotel as well as the 46th birthday of proprietor William T. Beasley in October 1908. This postcard was sent by Pearle Eade to her father, King City pioneer William J. Longacre. She wrote in part, "Dear Papa, Does this look like any place where you have been?" (Photograph by C. Harlan, courtesy Allen King.)

King City Rochdale Co. This business on the northwest corner of Broadway and Second Street was organized in January 1903. The trustees were Martin Kochesperger, Lyman B. Ulrey, A.E. Reynolds, B.F. Norris, and W.G. Curran, with J.E. Chope as manager. The store opened in the former Cooper Ulrey Company building, which the firm leased for five years at $50 per month. In February 1910, it was sold to A. Widemann Co.

Inside Rochdale Company Store. Formerly the Cooper Ulrey Company, this building is shown here in 1908. It is a typical scene from early stores. On the left is Byron Norris, one of the first managers of the Rochdale. Norris was one of the five original trustees when the company was founded in 1903. Longtime clerk Julius Brunetti is on the right.

A. Widemann Co. At the northwest corner of Second and Broadway, this general merchandise firm from Gonzales replaced the K.C. Rochdale Co. in February 1910. Identified in the photograph are manager Clarence Walker at the corner of the building with Jack Wasson, Nessen Schmidt, Joe Reinhart, and John Bean (?). Sisters Fannie Bachman and Lizzie Luckett are on the right with Julius Brunetti. In 1916, Walker-Brunetti Company took over the hardware and grocery departments.

Inside A. Widemann Co. Store. Based on the stamped metal roof panels, this building is most likely the A. Widemann Co. Widemann enlarged the store after taking over the Rochdale Company in 1910. At left is Julius Brunetti, who clerked at the store for 29 years. In 1916, Brunetti and Clarence Walker took over the operation as Walker-Brunetti Company but sold a year later.

Monterey County Bank Branch. This King City location was at the northeast corner of Broadway and Third Street. Construction began in the fall of 1907. It opened in February 1908 with H.E. Wetzel as cashier. The main office was in Salinas. Well-known architect William Henry Weeks designed the 25-by-60-foot building.

Bank in the Street. Starting in 1908, Monterey County Bank had its home at the northeast corner of Broadway and Third Street. In 1917, the structure was picked up and moved into the street to continue operating while a more modern structure was built on the same lot. A similar building was constructed in downtown Gonzales.

INCORPORATION ELECTION. On January 31, 1911, the residents of King City voted to incorporate the town as "the City of King." Elected trustees were J.A. Carlson, G.P. Henry, A. Goldstein, Peter Morasci, and R.R. Diaz with the highest vote-getter, Carlson, elected mayor. Elected clerk was A.B. McReynolds, the treasurer was Joe Bonadelli, and W.G. Curran was marshal. This February 4, 1911, article came from the *King City Rustler*. (Courtesy Monterey County Free Library.)

INCORPORATION CARRIED, WITH EIGHTEEN MAJORITY.

Incorporation carried last Tuesday, the total vote being 144, the majority in favor being 18, so we now live in "The City of King." However, before the newly-elected government can take control the county supervisors must officially proclaim the result of the election, next week. At that time they will correct a clerical error, and we shall legally become "The City of King City" and the name for ordinary purposes will remain just plain King City, as heretofore.

If the new trustees elect as chairman the member securing the highest number of votes, popular Al Carlson will become "the mayor." The vote polled was as follows:

Any resident who finds fault with the above would "growl if he was going to be hung." The trustees are all progressive men who combine the civic pride to make of this a model little city, and the conservative element of character to not go any faster toward this end than a fair and unburdensome taxation warrants.

For marshal Will Curran demonstrated a tremendous popularity, by being elected with a whopping majority after public notification through the press that he had withdrawn. It seems when his friends read of his withdrawal they sought him out and induced a change of mind on his part.

WASSON BLACKSMITH INTERIOR. Built in 1906 just east of 223 Broadway, this building adjoined the old Winckler store. It was constructed for Joe Wasson, who a year before came to King City from the stables at Hotel Del Monte in Monterey. This interior view shows, from left to right, Bill Wasson, unidentified, Jim Wasson in the back, and Joe Wasson. With blacksmithing in their blood, brother Bill continued in the business and was still shoeing horses in the 1970s.

First National Bank. The prominent building on the southwest corner of Broadway and Third Street was built in 1917. The opening was celebrated in November with a barbecue at the King

City Garage down the street. Lyman B. Ulrey, who owned the garage, was also the bank president. It later became Bank of Italy in 1921 and Bank of America in 1931.

Bank of Italy. This rare image of Broadway at Third Street shows the era after July 1921 when First National Bank of King City became a branch of the Bank of Italy, which had 34 branches in California. In 1930, this distinguished building became the Bank of America. In the distance on the left is the King City Hotel, which opened in January 1925, conducted by Walter W. Lundbeck.

Pacific Greyhound. A bus is parked in the alley alongside Yates Hotel and Soda Fountain at 123 Broadway mid-block between First and Second Streets. The hotel was formerly Brunetti's French Hotel until it was purchased by Abraham Yates in 1919. In 1921, son Bill Yates moved his fountain business here from the Hables building on the north side of Broadway. He celebrated the occasion by having a dance in his new hall at the Yates Hotel.

Bill Yates Confectioner. This important northwest corner of First and Broadway shows the newly constructed three-story Hotel El Camino Real, a concrete annex that included 40 rooms on the upper floors. The first floor had a store, dining room, kitchen, pantry, bakeshop, and butcher shop. To the left is C.R. Roberts Jewelry store and Bill Yates Confectioner. In 1921, Yates moved across Broadway to his newly purchased Brunetti Hotel and renamed it Yates Hotel.

Inside Yates Soda Fountain. The scene on this postcard is estimated to be in 1919 showing Bill Yates and Maye McCoey behind the counter. The soda fountain was between First and Second Street on the north side of Broadway, just west of the Hotel El Camino Real. (Courtesy Howard P. Strohn.)

First City Hall and Library. The small building sandwiched between the Brunetti Hotel in the center of this photograph and the IOOF Hall on the left was the first location of city hall after the town was incorporated in 1911. W.B. Weller rented the building to the city for $10 a month. This photograph from 1946 shows the Buttercrisp Bakery on the right, on the south side of the 100 block of Broadway. (Courtesy Mayo Hayes O'Donnell Library, Monterey History and Art Association.)

First Permanent City Hall. In April 1916, the city purchased the old grammar school building and grounds at 212 South Vanderhurst Avenue for $2,500. In July, city trustees met in the new quarters for the first time. E.R. Waters was hired as a janitor at $5 per month. In 1939, this building was moved to the northeast corner of Division and Canal Streets and used for a Boy Scout hall, leaving the Vanderhurst lot to house the new city hall.

City Hall. This building at 212 South Vanderhurst was a Works Progress Administration project during the Great Depression. Well-known Monterey architect Robert Stanton was the same man who designed the town's Robert Stanton Theatre on North Mildred Avenue. The dedication ceremony was held in August 1939. The library occupied the south end of the building, and over the years, it was shared by the health clinic, the California Highway Patrol, and the justice court.

Aerial View, c. 1920. Here is what Broadway looked like in the early days. The circle on the left at 404 Broadway appears to be a train track for children. The white house next to it belonged to Father P.A. Grammen. It later became the American Legion hall and was bought by Walter Foor and converted into his mortuary. In the distance are the railroad tracks and water tank.

Street Scene, 300 Block Broadway. In the days before asphalt, Broadway was the muddy main street of town. At far left is the post office building, which operated from 1892 through 1913. The H-A-F Company store on the corner was owned in part by W.J. Hamilton, the "H" in the name. Also seen is Monterey County Bank, which opened in 1908 at the northeast corner of Third Street.

Children's Parade. The vehicles and buildings in this photograph indicate that it was around 1917, with the participants heading west on Broadway in the 300 block near the intersection of Third Street. The question remains: What were they celebrating? Also, what is piled in front of the H-A-F Company store? By the shadows, it seems to be late afternoon.

Future Theatre Home. The ground-breaking ceremony of the new Reel Joy Theatre took place in March 1924. In the center is special guest and child star Jackie Coogan. The location is on the south side of Broadway, mid-200 block. Among the day's activities was a "fat man's" foot race down Broadway. It included Mr. Beasley, Mr. Henry, Mr. Burns, and Mr. Layous, with Layous winning. (Courtesy Monterey County Agricultural and Rural Life Museum.)

Recently Built Reel Joy. This two-story theater building was constructed for owners "Ma" and "Pa" Hables, featuring rooms above and two storefronts on street level. It opened on February 3, 1925, with Buster Keaton starring in *The Navigator.* In March 1925, it hosted the premiere of *Burning Trail*, starring William Desmond and filmed in the Jolon area. In 1929, "talkie" movies were introduced.

Snow, 1922. This rare snowfall was enjoyed by children seen next to Broadway near Vanderhurst Avenue. The prominent structure in the background was known as the Father Grammen house. It was purchased by the American Legion in 1921 for their clubhouse. Walter Foor, the local mortician, acquired it in 1924 and remodeled it to accommodate his business. Today, it sits at the back of the lot and faces Lynn Street. (Courtesy Dughi family.)

Snow Scene Downtown, 1950s. Snowfall in King City is an unusual event, which makes this view of the intersection of Third Street and Broadway rare. Some old-timers remember students at the local high school throwing snowballs that year, but with bad aim, so several windows suffered. School administrators asked students over the public address system to please stop throwing snowballs. (Courtesy Roger Elzey.)

Awaiting the Liberty Bond Train. In September 1918, an estimated 3,000 people greeted the arrival of the train when it stopped in King City. All businesses were closed as the Boy Scouts with their drum corps led the parade followed by school children carrying flags. Two open flatbed cars carried exhibits of cannons along with machine guns and so-called trophies captured from the Germans. This photograph was taken from atop the Southern Pacific Railroad water tank.

Water Tower, Built 1929. Resembling the Tin Man from the *Wizard of Oz*, this King City Concrete Pipe Company photograph shows the 100,000-gallon steel tank that served the town for many years. It was 100 feet high and replaced a 50,000-gallon wooden tank with the hope insurance rates would come down with more storage and pressure. Located in the Pacific Gas and Electric yard on North Second Street, it was dismantled in June 2021.

Truck Load of Lumber. In this photograph dated January 10, 1928, building materials are ready to be delivered from the S.P. Milling Company at the northwest corner of Lynn and First Street. George Hays ordered the lumber for his boardinghouse to be constructed on Lynn Street. In the back to the east, next to the railroad tracks, is the Southern Pacific Grain Warehouse. (Courtesy Jernigan Collection, Carol Smith Sundahl.)

DOCTORS, NURSES PRETTY TIRED

Flu Is Here, but Town Sits Straddle of It and Holds It Down

The Spanish influenza is here. It sneaked in like a Hun, laying several of our people low, but quick action by the city in passing a mask ordinance and the prompt steps taken by the Home Relief Section of the Red Cross has seemingly prevented its spread.

Spanish Flu. Written in 1918, this headline could have been used for the COVID-19 epidemic 100 years later. During the Spanish flu, a temporary hospital was set up at King City High School, and the article mentions that residents were "worn out from volunteer work" and there was a need for more volunteers. As this book was compiled in 2021 and 2022, King City's medical providers were working under similar weariness. This article was published in the *Rustler* on November 1, 1918. (Courtesy California Digital Newspaper Collection.)

Six

THE PROGRESS OF EDUCATION

Providing education for children was a topic of discussion as soon as the town began in 1886.

An early Sanborn map lists "School" on a building on the north side of the 200 block of Broadway. (Sanborn maps, used for insurance purposes, are among the earliest resources documenting buildings in many towns.)

Records from the Monterey County Office of Education list early school teachers as Rosalie Branstetter and Olive Hamilton, John Steinbeck's mother. She taught during the fall 1889 and spring 1890 terms.

The city founder, Charles King, donated land for a permanent grammar school on the 200 block of South Vanderhurst Avenue, which later became the site of city hall.

Then in November 1910, Monterey County voters approved the establishment of two county high schools: one in Monterey and one in King City. King City High School (KCHS) students first met upstairs in the Talbott building at 321 Broadway. In November 1911, the board of education purchased land in what was known as the Hersom Tract on the north side of the 700 block of Broadway. Work began in March 1913 on the new high school and students were in the structure by November.

At about the same time, the King City Grammar School was expanding. Property on Pearl Street was secured, and $20,000 in bonds were sold to First National Bank for a new campus at Pearl Street and Vanderhurst Avenue in November 1911.

Both schools were designed by well-known architect William Henry Weeks to be stately and similar in appearance to the twin towers.

The treasure of the King City school buildings is the King City High School auditorium. Erected in 1937–1939, later known as the Robert Stanton Theatre, it was located on the high school campus on the 100 block of North Mildred Avenue. A product of the Works Progress Administration, it featured concrete construction designed by architect Stanton and Art Deco sculptures by well-known artist Jo Mora.

Over the years, other school campuses built were Santa Lucia, Del Rey, Chalone Peaks Middle School, St. John's School (a Catholic institution), and Hartnell College (formerly Salinas Junior College). The original King City Grammar School went on to be called San Lorenzo School and the Arts Magnet School. Students educated in these institutions have gone all over the world and influenced others.

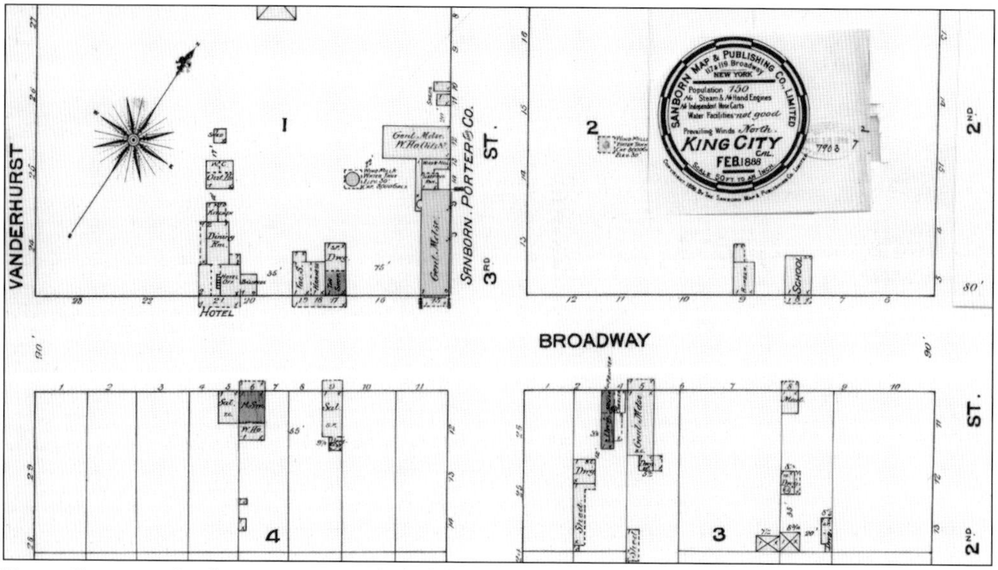

First School. Sanborn maps are helpful in showing the locations of early buildings. This February 1888 map shows a school and Tom Himmah's barbershop at upper right on the north side of Broadway between Second and Third Streets. It was here that Pearl Brown, C.H. King's sister-in-law, held a private school for a reported four months. In the fall of 1888, a permanent structure was completed, and Rosalie Branstetter started teaching at the present city hall site.

First Permanent Public School. Six lots were deeded in March 1890 by C.H. and Kate King to the King City School District at this location on the northwest side of Vanderhurst Avenue between Bassett and Pearl Streets, the present-day city hall location. Trustees were Adolph Winckler, Eugene Neuman, and John Phillips. Among the first teachers in King City were Roselie Branstetter (1888) and Olive Hamilton (1889).

Grammar School. Students began attending classes here in the fall of 1912 on the south side of Pearl Street. The front of the school faced Vanderhurst Avenue, just across the street from the old school (on the city hall site). In the *Rustler*, seventh-grade pupil Bertha Decker described it as "a large concrete building built in the Mission style." Its two towers were noted to be a twin of the high school.

Grammar School Auditorium. The two-story building on the right is the San Lorenzo Grammar School Auditorium facing Pearl Street. The classroom building on the left faces South Vanderhurst Avenue. It was built in 1912, but the auditorium was not constructed until 1921. The auditorium was formally opened on June 3, 1921, with graduation exercises. The following week saw its first social event when the American Legion sponsored a dance.

FIRST HIGH SCHOOL, 321 BROADWAY. This two-story building with bay windows was the Talbott Building. It was constructed for local county supervisor Paul Talbott in 1908 for rental purposes. The first occupant was the *Rustler* newspaper offices on the ground floor. In 1911, the first high school class met in the upper level because no school facilities were available. By 1923, the upper level served as King City's first hospital.

HIGH SCHOOL. Constructed in 1913, this distinguished building graced the corner of North Mildred Avenue at 720 Broadway with students first attending in November. It was originally a county high school under the county board of education. In 1915, the King City Union High School District was formed with 45 feeder schools covering all of southern Monterey County. The first trustees were Neil Forden, J.D. Dunovant, Julius A. Trescony, John L. Matthews, and E.W. Kirk.

Heavyweight Basketball Team, 1925. King City High School's varsity "A" basketball team and its coach pose beside the school. This was the year it won the league championship when it beat Pacific Grove by two points. The *Rustler* reported, "King City has turned out this year the strongest athletic teams since the high school started." Credited with helping win were Bunte, Imwalle, Russell, Luckett, Soberanes, and Captain Hamilton. (Courtesy Robert Mansfield.)

Girls' Basketball, 1921–1922. KCHS has a long record of supporting sports for girls. This photograph shows the 1921–1922 girls basketball team. In February 1922, faculty members Grace Maxwell and Caddie Cook entertained the team in their home. Present were Carol Bunte, Velma Campbell, Charlotte Bruce, Doris Morrison, Elsie Arfston, Nellie and Eva Carlson, Rilla and Edwina McReynolds, Marion Barbree, and Lurline Brunetti.

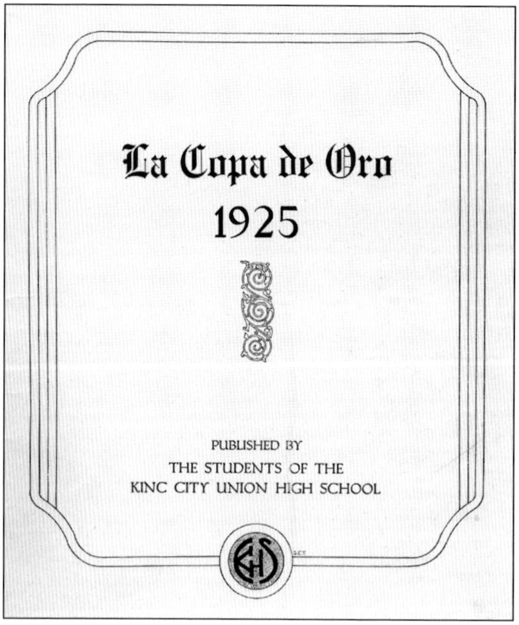

La Copa de Oro
1925

PUBLISHED BY
THE STUDENTS OF THE
KING CITY UNION HIGH SCHOOL

KCHS YEARBOOK, 1925. *La Copa de Oro* was the title of the King City High School yearbook produced in the early years, meaning "cup of gold" in Spanish. An interesting feature was the listing of past graduates by year. In 1913, the first graduate was Mary McCormack, who taught in the Long Valley School from August 1915 to June 1917. Later, she became Mrs. Barney Teague and was living in Los Angeles.

The Alumni

YEAR	NAME	OCCUPATION	ADDRESS
1913	Mary McCormack	Teaching	San Joaquin
1914	William Brinan	Army	Camp Kearney
	Florence Dudley (Mrs. C. Taylor)	Housewife	Parkfield
	Arthur Himmah	Ranching	King City
1915	Ruth Vivian	At home	King City
	Phyllis Henry	Business Col.	Santa Cruz
	Leona Hunter	At home	Salinas
	Gwendolynne Buttle (Mrs. Stonesifer)	Housewife	Philadelphia
	Leila Currell	At home	King City
	Julia Garcia (Mrs. Allen Jacks)	Housewife	King City
	Craig Mather	Hos. Corps	San Francisco
	James Wasson	Army	"Over There"
	Frank Condon	Storekeeper	Salinas
1916	Lepha Sitton	At home	Arizona
	Beatrice Vivian	Normal	San Jose
	Elmer Forden	Army	"Over There"
	Carol Marcus (Mrs. Sharp)	Housewife	Berkeley
	Alta Ulrey	University	Berkeley
	Thomas Brinan	Ranching	San Ardo
	Dorothy Weidemann	At home	Bradley
	Ruth Branstetter	Normal	San Jose
	Miriam Whitney	At home	San Ardo
	Muriel Bunte	Tr. for nurse	Santa Cruz
	Ada Nelson	At home	Greenfield
	Lucille Kirk	At home	Jolon
1917	Floyd Wimberly	Stan. Univ.	Palo Alto
	Clyde Bushnell	Navy	San Pedro
	Marie Griffin	University	Berkeley
	Adele Johnson	Book-keeper	King City
	Helene Veach	Book-keeper	King City
	Nellie Weaver	University	Berkeley
	Ethel Page	Normal	San Jose
	Vernie Davis	"Hello Girl"	King City
	George Brinan	Ranching	San Ardo
	Ellen Dani (Mrs. F. Zoellin)	Housewife	King City
	Ursula Whitney	At home	San Ardo
	Robert Griffin	Ranching	King City

—GRACE BUSHNELL, '19.

AUDITORIUM UNDER CONSTRUCTION. Here is a view of the King City High School auditorium on North Mildred Avenue with scaffolding still in place. On October 19, 1936, the school district held an election for approval to spend $48,000, which trustees had budgeted as their share of the project. The overall anticipated cost was $120,000. The building was completed in 1939, a Works Progress Administration project. (Courtesy Monterey Area Architectural Resources Archive.)

HIGH SCHOOL AUDITORIUM. The Robert Stanton Theatre is named after the architect of the Works Progress Administration project and was built between 1937 and 1939. Originally called the King City High School Auditorium, it is located on the 100 block of North Mildred Avenue. It was planned to be built facing Broadway, but parking and traffic concerns changed the location. The building was listed in the National Register of Historic Places on July 23, 1991.

Jo Mora Design. The King City High School Auditorium, now home to the Robert Stanton Theatre, features this bas-relief architectural work on the building facing North Mildred Avenue. A native of Uruguay, Mora eventually settled in Monterey County where his artistic creations became well known. The panels express distinct times, with the three lower panels showing Native Americans, the center panels the Renaissance period, and the upper ones reflecting modern American history.

KCHS Auditorium. The interior of the King City High School Auditorium has been the scene of many theatrical productions such as this one. Starting in 1947, the King City Lions Club presented an annual all-male show. The theme in 1950 was "Streets of Paris." Could this photograph be a scene from that production? The building, with its splendid acoustics, was renamed the Robert Stanton Theatre in February 1990. (Courtesy Houx family.)

Patterson Grove. This was the view from the roof of the King City High School Auditorium shortly after it was built in 1939. North Mildred Avenue is in the foreground near where it meets Ellis Street. Behind that is Patterson Grove, a cluster of eucalyptus trees owned and cared for by Ellis E. Patterson, who purchased the grove from Dr. Dorus Brumwell.

KCHS Campus Aerial View. The oval on the left is War Memorial Stadium on the King City High School campus. Built in 1948, it opened in October with a football game against rival Gonzales Spartans. The King City Lions Club spearheaded the drive to dedicate it in September 1952 to those who lost their lives in World War I, World War II, and the Korean conflict.

KCHS School Buses. Soon after King City High School was built in 1913, these buses lined up in front of the campus. Bus routes included Greenfield, Arroyo Seco, Lockwood, Metz, and San Ardo. Although Bitterwater is in San Benito County, it was in the King City High School district with a bus run. Early on, students took the train to and from towns south of King City. (Courtesy Monterey County Agricultural and Rural Life Museum.)

Hartnell College Bus. In March 1971, this vehicle transported King City area students to the junior college campus in Salinas. Because of this need, long-time Hartnell College supporter Villeroy "Roy" Gleason and his wife, Phyllis, bequeathed their home at 404 Ellis Street to the Hartnell College Foundation. It became the Gleason Center, where classes were held. Needing more space and an updated facility, the present campus was built in 2002 at Second and Lynn Streets.

Seven

Early Agriculture and Irrigation

The story of King City follows a journey that has transformed it from the great Salinas Desert to a productive and fertile valley.

Long before it was referred to as the "Salad Bowl of the World," the region around King City was home to cattle, sheep, hogs, and dryland farming.

In May 1861, land surveyor William H. Brewer recorded in his diary that "with water, this would be finer than the Rhine Valley itself; as it is, it is half desert." How prophetic his words were.

Today, irrigation has made the area around King City a beautiful mosaic of precision farming with fields containing multiple colors of lettuce mixed with spinach, broccoli, cauliflower, and other crops. To convert that desert, the area needed irrigation, which was brought first by canals and later by pumping groundwater.

In 1897, after Charles King sold 12,000 acres of Rancho San Lorenzo to Spreckels Sugar Company, more effort was made to produce sugar beets, beans, and other crops using a sharecropper system in which farmers shared a percentage of their crop with the landowner. The land was also divided into dairies and sold to mostly Swiss and Italians.

The Salinas Valley was discovered to be a good home for sugar beets, pink beans, and lettuce. Carl Maggio opened a carrot shed. Leo and Bob Meyer opened a tomato packing plant, as did Brown and Hill (later Frudden Produce).

In 1957, after the Mesa del Rey Flying School closed, its property on the bluff overlooking the east side of the city was purchased by Basic American Foods for onion and garlic dehydrating. People used to say they knew they were home when the smell of garlic and onions greeted them as they drove over the Salinas River Bridge or when they approached Wildhorse Canyon from the south.

The influence of fieldworkers from Mexico helped shape the town. These hard-working people brought food to our tables and are loyal to local businesses. Many spend their earnings at markets that cater to goods and foods from Mexico. The Bracero program brought in workers. In more recent years, the H-2A worker program has provided labor to harvest local fields.

The lyrics of a song from 1969 by John Stewart called "Mother Country" aptly describes the farmers, fieldworkers, and King City business workers: "They were just a whole lot of people doing the best they could."

HAULING HAY WITH 10 HORSES. This 1880s photograph shows Garnet Herbert's horse team pulling three wagon loads of hay. Behind the horses are the Southern Pacific Milling Co. warehouse, the Southern Pacific Railroad depot, and the livestock corrals with a livestock train car. The Herberts lived across the Salinas River from King City and, early on, raised barley on the Dunphy property, which later was acquired by the Salinas Land Company.

GRAIN HARVESTING. Here is a typical harvesting operation with a track layer tractor pulling a grain harvester. Throughout southern Monterey County, thousands of acres were planted in barley or wheat. Known as dryland farming, these crops were dependent on rain to provide moisture. When the price dropped and it became unprofitable, much of the land was returned to grazing or replaced by grapes. (Courtesy Dan and Celie Hayes family.)

Sugar Beet Harvest. Here is a typical harvesting scene from when the sugar beet was king in the 1940s and 1950s. On the left is a "spike harvester." Beets were loaded into a truck and then hauled to the beet dump at the railroad tracks to be loaded into open train cars and transported to the factory for processing. Note the beets stuck on the spikes as they are picked up from the field. (Courtesy Dan and Celie Hayes family.)

Hauling Sugar Beets. Spreckels Sugar Company's beet dump at Elsa is about three miles north of King City near the King Ranch headquarters. Here, eight-horse teams pull the loaded wagons for the beets to be loaded into open train cars to be shipped to the sugar factory at Spreckels. In 1904, it was noted that a ton of sugar beets would yield 210 pounds of refined sugar. (Courtesy Monterey County Agricultural and Rural Life Museum.)

SPRECKELS PUMPING STATION. Built in 1897–1898, this facility provided irrigation for Spreckels Sugar Company's Ranch No. 3 at King City. The water was drawn from a sump pit near the river to tap the underground stream. In the beginning, three centrifugal pumps were turned by three steam engines propelled by eight boilers. In this photograph, the exhaust stacks from the boilers tower above the building. (Courtesy Monterey County Agricultural and Rural Life Museum.)

STEAM PLOUGH. This postcard, dated 1907, shows a strange-looking piece of equipment on the Spreckels Ranch No. 3, northwest of King City. The tall stacks in the background are the site of the Spreckels pumping plant behind the cemetery. Note the cable from the plow to a spool on the belly of the steam-powered tractor. Blacksmith John Deere is reported to have made the first steel plow in 1837. This photograph was taken by James E. Meddaugh. (Courtesy Allen King).

King City Concrete Pipe Co. This business helped transform the valley with irrigation in the 1920s–1940s. The manufacturing yard was northwest of Ellis and First Streets and was run by a partnership of C.I. Ragus, John L. Matthews, and Mike Corak. Some of this concrete pipe ended up on the San Bernabe Ranch south of King City, along the Jolon Road, and on the Salinas Land Company north of King City.

Eucalyptus Trees. Windbreaks were part of the viewshed in the Salinas Valley for decades. To protect crops from the daily west winds, rows of eucalyptus trees were planted at intervals in the fields. In the 1920s, the Salinas Land Company, northwest of King City, constructed a nursery and raised 170,000 trees, with many going to windbreaks. Today, they are no longer visible from Highway 101 because they were removed. (Courtesy Salinas Land Company and California Orchard Company.)

Early Irrigation Example. The Garnet Herbert family worked to water row crop land west of King City. Before sprinklers came into use, irrigation pipe was laid and water ran down the furrows. Through the years, crops have included sugar beets, beans, tomatoes, carrots, potatoes, lettuce, broccoli, and leafy greens. Pictured here are, from left to right, Walter, Garnet Sr., David (Walter's son), and Garnet Jr. (Courtesy Salinas Land Company and California Orchard Company.)

Cattle at "Cow Camp." Brothers Les Smart (left) and Bill Smart are shown here in the foothills west of King City checking their herd of Hereford cattle. The cattle industry has long been an important part of southern Monterey County. In the 1920s, brand inspector R.R. Diaz reported more than 14,000 heads shipped in a year from Metz, King City, and San Lucas. (Courtesy Salinas Land Company and California Orchard Company.)

Charles Lombardi Dairy. When the former Spreckels Sugar Company holdings were broken up by Chris Widemann, the land use changed from sugar beets and other crops to alfalfa and dairies. In the foreground of this photograph is the Alpine Milk Company condensery at Coburn, north of King City, a direct result of dairies being established. Starting in the 1910s, many Swiss-Italian families settled in the area off Metz Road next to the railroad tracks. (Courtesy Steve Lombardi.)

Alpine Milk, South First Street. This milk plant was between San Lorenzo Creek and Lonoak Road east of the railroad tracks. L.F. Roussey, the owner of the property, deeded approximately 12 acres for the condensery and construction started in March 1927. In 1932, Alpine Milk Company became Nestle's Food Company. Years later, it became home to the Maggio carrot operation. As of 2022, it is the location of Gill's Onions. (Courtesy Robert L. Meyer.)

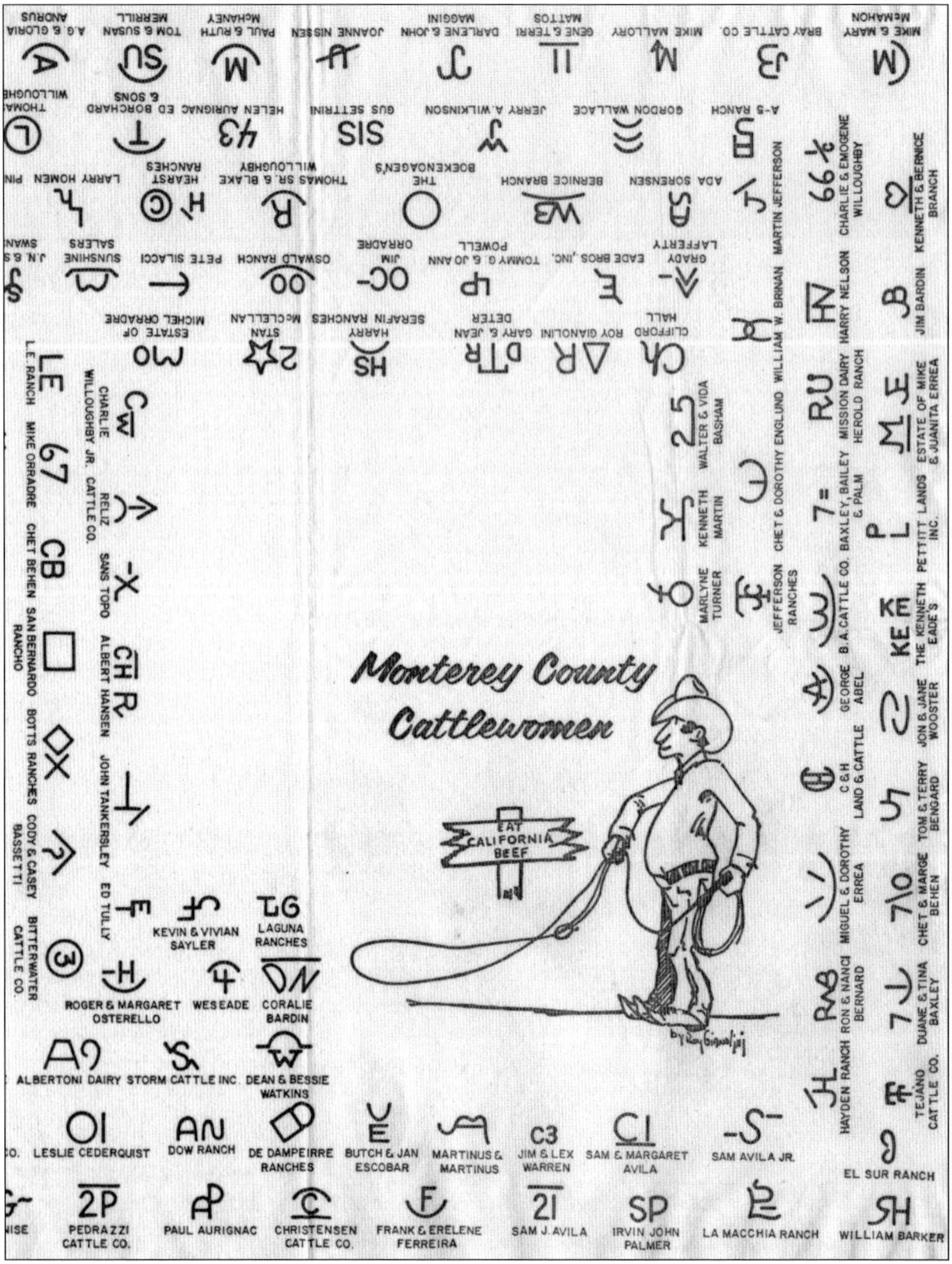

CATTLE BRANDS. Here are some samples of historic brands registered to residents of the King City area as featured on napkins used to promote the beef industry. Perhaps the first brand used where present-day King City is was registered by E. Sherwood on April 5, 1860, featuring an arrow pointing upward. In addition to marking livestock, brands represent the hallmark of a ranching family and also symbolize self-esteem, trust, and stewardship. Well-known rancher Roy Gianolini of Greenfield contributed his artistic talent in the cowboy drawing. (Courtesy Monterey County Cattlewomen.)

King City Stampede, 1936. Here is the scene of the first Stampede (rodeo) in Hog Canyon, three miles east of King City off Bitterwater Road. Ranchers and twins Jess and Warren Cornwall hosted it on their property as part of King City's Golden Jubilee. Its success prompted the community to continue it for many years. Note the Gabilan Mountain Range in the background to the east.

Loading Pink Beans. In the 1940s, King City pink beans are being unloaded by Larry Hearne, who founded King City Warehouse at 1023 Broadway. It later became L.A. Hearne Co. and moved to 524 Metz Road. When it was "king," some growers planted up to 60 percent of their tillable land in pink bean. (Courtesy Salinas Land Company and California Orchard Company.)

CARROT HARVEST. This scene shows early carrot harvesting, perhaps in the 1930s. The tractor at left with iron wheels helps identify the era. Note the girl standing on the right. Is she there with her parents or one of the laborers? Foreign workers have a long history of helping get crops from fields to store shelves. (Courtesy Monterey County Historical Society/Pat Hathaway Collection.)

HARVESTING LETTUCE. This early photograph of lettuce harvesting shows the innovative nature of farmers as this equipment appears to be homemade. Using the chassis and engine of a vehicle, they added a bin and conveyor belt to make the operation more efficient. Due to the early success of growing lettuce in the Salinas Valley, it became known as "Green Gold." (Courtesy Monterey County Historical Society/Pat Hathaway Collection.)

Harvesting Table Grapes. In 1929, the California Orchard Co. west of King City grew grapes that were packed into their COCO-brand boxes. The first planting of grapes was at the missions. In 1919, the first planting of Salinas Valley commercial wine grapes was begun at Chalone Vineyards, east of Soledad. During the "grape rush" in the 1960s and 1970s, the industry came into its own. (Courtesy Salinas Land Company and California Orchard Company.)

World's Largest Salad. Members of the King City Young Farmers organization stand behind their "World's Largest Salad." A publicity stunt they staged at the 1982 Salinas Valley Fair in King City recognized the Salinas Valley as the "Salad Bowl of the World." On the right, a list of supporters from the produce industry can be seen. Some attendees are holding bottles of salad dressing. (Courtesy Pozzi's Stampede.)

MAGGIO CARROTS. The Carl Joseph Maggio packing complex was south of San Lorenzo Creek on First Street near Lonoak Road. It began its packer-shipper operation in 1952. Among upgrades was the establishment of a vacuum packing cooling division. The operation became the largest grower-shipper in the nation for carrots, lettuce, and bell peppers. It was originally the home to Nestle's milk plant and more recently Gills Onion's.

WOMEN IN AGRICULTURE. Pat Gill, left, and Pat Anthony were two entrepreneurial King City women married to farmers. Their unique agricultural contribution was introducing the pimiento, a sweet red pepper, as a fresh market product. Established in 1979, Pat's Pimientos marketed and sold fresh pimientos grown in the Salinas Valley to gourmet and consumer fresh markets. The business expanded into the fancy food industry with pimiento jam and relish products. (Courtesy Gill family.)

Eight

A Highway Runs through It

The influence of US Highway 101 is pivotal to the story of King City.

The early routes into King City were dirt roads and bridges used by horse and wagons and stagecoaches. Early state routes followed the Salinas Valley from Soledad south through Jolon, the path known as El Camino Real that connected the missions.

With the development of the automobile, California legislators saw the need to develop a continuous route from north to south through the state. Starting in 1912, two options were considered by state leaders: one followed the Jolon route, and the other stayed in the Salinas Valley. Residents of Jolon and King City lobbied for their towns, and King City won.

In this chapter, readers are asked to envision starting at the south end of First Street and experience what travelers would find entering from the south: gas stations on almost every corner, vehicle repair shops, restaurants, hotels, and auto dealerships. Patrons of downtown businesses regularly dodged speeding vehicles.

In the 1960s, the state began discussions to bypass downtown King City with a highway. Again, local debate polarized those favoring a route on the west side of the river and those wanting the road on the east side, closer to King City businesses. The east side route, known as the K-line bypass, was approved. In 1968, the route through downtown stopped. Popular eating establishments such as Keefer's Restaurant moved to the new highway. However, this began a slow decline of the downtown, which happened across the country when highways relocated.

In the early 1970s, a photographer produced a report called "A Stranger Came to Town," visually showing some of the rundown effects when the money from thousands of cars per day no longer traveled the streets of a city.

Gone today along Broadway are the types of businesses from the Highway 101 era. Many of the buildings have been successfully repurposed to create the charm and new services the town has today, including a preponderance of restaurants featuring wonderful Mexican food. Several historic buildings in the 200 block of Broadway were demolished, transforming the block into the more modern Town Square. Seasonal banners along the main street proclaim the route as Historic Highway 101.

The taxes from gasoline sold in King City have greatly benefited the town over the years.

KING CITY GARDENS. Marian Doyle started her floral and nursery work in a harness shed on Spreckels Sugar Company Ranch No. 3 north of King City in 1928. Four years later, a store, greenhouse, and her family's home were constructed at 702 South First Street, shown here. Her husband, Frank, ran the Mission Trails service station that they built nearby. In 1944, their daughter Frances and son John (and his wife, Marie) entered the business. (Courtesy William T. "Bill" Doyle.)

KING CITY AUTO GLASS. The southwest corner of First and Pearl Streets, shown here in March 1961, was typical of the auto repair businesses that lined the entryway into King City. The building was erected in May 1925 for Earl Olds, who moved his garage and machine shop from the 100 block of Broadway, where he was selling Hudson and Chevrolet cars. An ad showed Olds carried Havoline oil and Navy gasoline.

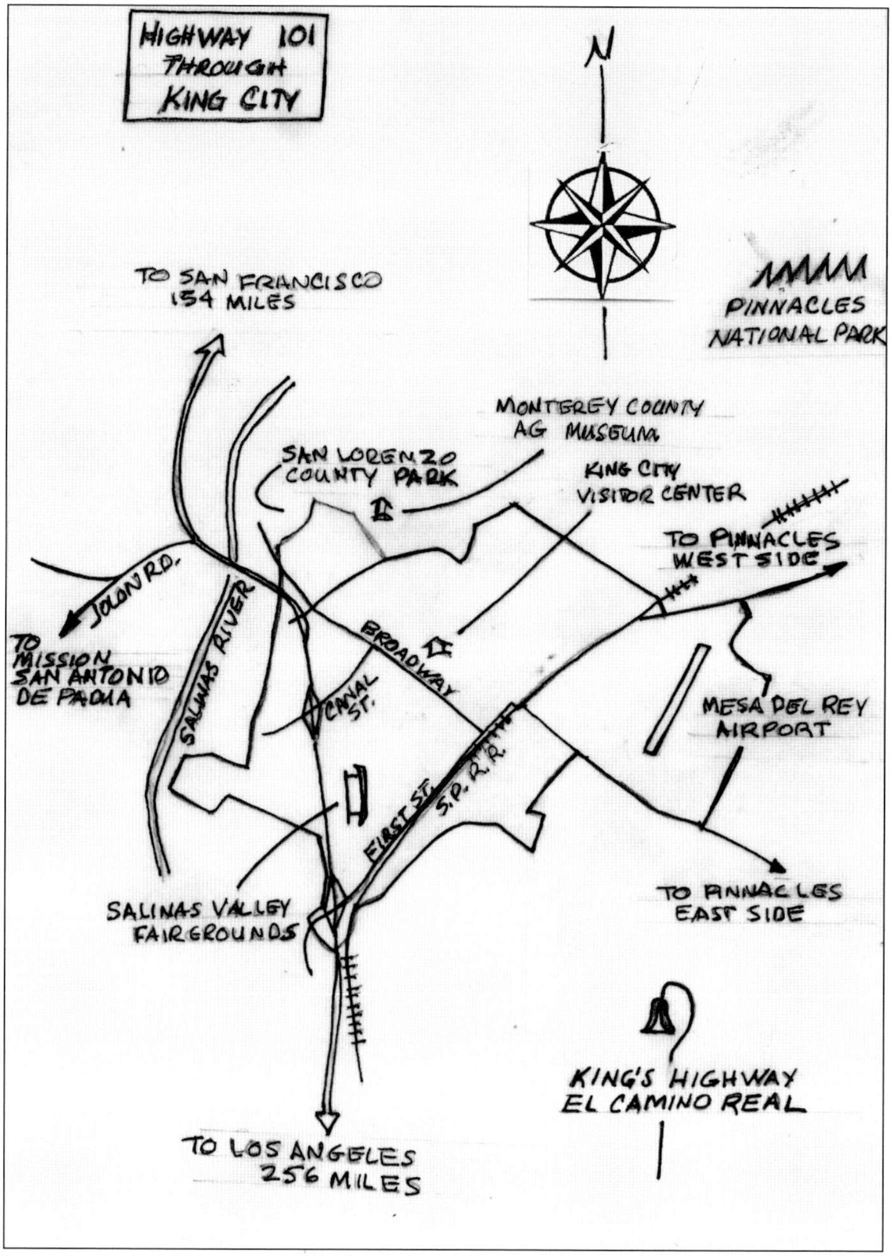

US Highway 101 Map. King City's location about midway between San Francisco and Los Angeles has made it a frequent stopping place for travelers. With an hour's drive between the bigger towns of Paso Robles to the south and Salinas to the north, King City has offered services to many travelers, including access to the only hospital between the two cities. King City is also on the route to the east and west sides of Pinnacles National Park by way of Metz Road and Bitterwater Road as well as the Jolon Road route to Fort Hunter Liggett, Mission San Antonio de Padua, Lake San Antonio, and the Los Padres National Forest. Highway 101 makes a swooping curve when it reaches King City and crosses the Salinas River. Beware of speeding when passing King City, as the California Highway Patrol has a reputation for keeping drivers safe by enforcing speed limits. (Map by Chuck Krause.)

"The Curve." That's how the intersection at First Street and Broadway was long known. Originally a 90-degree right-angle turn, it became more dangerous as traffic and speed increased. The Hotel El Camino Real was host to many big rigs laying on their sides. In 1936, the service station was moved back from the left side, and a curve was constructed in a Works Progress Administration project. (Courtesy Monterey County Agricultural and Rural Life Museum.)

Broadway Accident. A big rig rolled over on the sidewalk in front of the hotel and El Camino Motor Company. This became more frequent as truck traffic on Highway 101 increased. In 1935, the city bought a portion of the southwest corner, and the present curved turn was constructed. (Courtesy Karen Meurer Park.)

Pozzi's Stampede. This bar and restaurant at 207 Broadway has been a fixture for more than 80 years. It holds one of the oldest liquor licenses in Monterey County that allows both on-site and off-site sales. Pictured here at center are Mary and Emilio Pozzi, who founded it in 1934. Descendants of Swiss immigrants, they were well known for their polenta and stew. Today, it is run by the next generation of Pozzis. (Courtesy Emilio Pozzi family.)

Mar's Café. The Mar family came to King City in 1928 and bought the store of the late Mah Lon Sing, thus continuing the Asian influence. Sing opened his store before 1900. Asians were here before King City was laid out and they played an important part in building the railroad. Young Mar bought this building at 211 Broadway in 1938 and eventually opened Mar's Café. It became Lois' Café and then Guadalajara Restaurant, featuring Mexican food.

BUSINESS DISTRICT, THIRD AND BROADWAY. Here is a glimpse of the town's main street looking east. This photograph from a postcard is dated around 1938 based on the movie name on the marquee playing at the Reel Joy Theatre: *Having a Wonderful Time* with Ginger Rogers. Note

the sign in the middle of the street that reads, "Stop when pedestrians are in crosswalk," to help ensure the safety of residents when Broadway was the thoroughfare for Highway 101.

May Day Downtown. This celebration featuring a May Day pole looks as if it dates to 1954. The location is at the intersection of Broadway and Third Street showing the famed map near the side of the City Café building. In the 1950s, Broadway was still US Highway 101. Apparently, a special permit was granted so traffic moved to another street. Note that all the participants are girls.

Street Maintenance on Broadway. Road improvements around 1920 used different equipment than in modern times. Here, work is being done in the middle of the 300 block of Broadway with a view of Shaw's Vulcanizing Works and King City Battery Company at left on the north side of the street. In the mid-1930s, the site became home to Suk's Tavern and later the well-known Keefer's Restaurant.

Home of Pink Bean Soup. Suk's Tavern at 322 Broadway professed to be the best place to eat in King City after it was opened in 1934 by Anton Suk, a native of Czechoslovakia. Suk planned it to resemble an English house. The outside and inside murals were painted by Alfred Dupont of New York City, who also did murals at Hearst Castle. King City pink beans were a favorite of travelers and residents.

KEEFER'S PINK BEAN SOUP

- 3 c. (1 lb.) pink beans, rinsed
- 4-5 slices bacon, diced
- 3/4 c. diced celery
- 3/4 c. diced carrots
- 1 c. diced onion
- 1 clove garlic, minced
- 10-12 ozs. ham hock
- 1/2 c. tomato sauce
- 8 c. water
- Salt and pepper to taste

Cover beans with cold water and soak overnight. In a large pot, sauté bacon until crisp. Add celery, carrots, onion and garlic, and cook 5-10 minutes until softened. Drain beans and add to vegetables along with ham hock, tomato sauce and water. Bring to a boil, then simmer about 2-1/2 to 3 hours until beans are soft. Take out ham hock, remove and chop the meat from the bone, and return meat to the soup. Remove 1/3 of the beans and mash them, then return to the pot, stirring well. Simmer 20-30 minutes, stirring occasionally. Skim off any grease and season with salt and pepper.

Tim Keefer

Keefer's Pink Bean Soup. This is perhaps the most popular and well-known dish served in King City. When Jerry Keefer became the head chef at Suk's Tavern in 1947, the story went that owner Anton Suk told him, "Make something new, maybe pink bean soup." Keefer's reply was "Mr. Suk, I'll make that soup any color you want." Jerry's son Tim Keefer provided this recipe for the *Mee Memorial Hospital Service League Cookbook* in 2000.

King City Garage. This building at 315 Broadway was expanded in 1917 from a 50-foot front to 100 feet by owner John L. Matthews, reported in the *Rustler* as the "largest in the county." Originally built in 1910, it was home to Eatondale Garage, purchased by L.B. Ulrey and renamed King City Garage. Over the years, it housed numerous auto dealerships and later Tri's Foodliner, Fairway Mart, and El Pueblo Market.

Los Ositos Garage. One of the largest buildings downtown at 401 Broadway was built for Paul Talbott. From 1917 to 1919, it was the Los Ositos Garage, but over the years it has housed businesses such as Chaney's Super-Service, Tavernetti Motors, Turner Chevrolet, Valley Motors, Ford's Department Store, Napa Auto Parts, and Hayashi and Wayland accountants. More recently, it has been the district office for the King City Union Elementary School District.

STREET SCENE, 1936. Here is a look at the intersection of North Vanderhurst Avenue and Broadway during the Golden Jubilee celebration. On the northeast corner is the Richfield service station, which replaced Vanderhurst Hall. In the back is the fourth home of the *Rustler*. The vacant northwest corner on the left is currently the Monterey County Free Library King City branch, formerly a Bank of America. (Courtesy Monterey County Agricultural and Rural Life Museum.)

PETER SYLVESTER STUDEBAKER. This dealership offered motor service and sales at 428 Broadway. Pictured around 1949, it was originally built for Charles Van Buren Jackson in 1928 to house the second location of the King City Garage. Other businesses through the years were Livingston Brothers Tractor Company, Harry Scott's Implement Agency, Wally Berry's Brake and Wheel Service, J&J Auto Supply, Dick Bruhn's (A Man's Store), and Town and Country Ladies Wear.

Union Oil Products with Blimp. This service station, pictured in 1932, was at the southwest corner of Russ and Broadway, now Chase Bank. In the back is the Ed Waters house, which W.A. Lake bought in 1945 for his King City Motors. This photograph is featured in a delightful self-published book called *Food for Thought* in which Margaret Harrell wrote short essays about her life in King City. (Courtesy Margaret Waters Harrell Collection, Monterey County Historical Society.)

La Hacienda Motor Inn. Imagine this building mid-block on the north side of the 600 block of Broadway. Originally built as a home for John L. Matthews and his wife, Rose, it was remodeled by Rose in early 1925 as a place for tourists to dine and stay overnight. Events and social gatherings were held here until mid-1927. In January 1932, Dr. L.M. Andrus leased it and opened his community hospital.

King City Hospital. Constructed in 1941 by Dr. L.M. Andrus, this was a 22-bed facility with the latest X-ray and laboratory equipment of the time. Located at 624 Broadway, it was behind the J.L. Matthews residence, which had been an earlier hospital. After Mee Memorial Hospital opened in 1962, Andrus converted his facility into the Pioneer Hacienda, a long-term care facility.

Palm Motel. Few towns can boast of a stately palm tree growing through the roof of the lobby of a motel. In the 1950s, the original owners of the Palm Motel treasured this historic and rare monkey coconut palm, so they built around it when they decided to construct a motel at 640 Broadway across from the high school. It was a sad day for many residents when this unique tree was cut down in August 2020.

Lindy Motor Inn. This site at 743 Broadway was originally part of the Talbott tract. Lots were purchased by Mae Hector of San Francisco in 1927. A feature of the inn was a light-up airplane with the name "Lindy," a reference to Charles Lindbergh. When Bert Kerns bought the place, it was renamed the Surrey Motel and featured a full-size horse and surrey.

Mount Implement Co. Shown here is the L.W. Menzies property at the west end of town across from the cemetery. In 1935, Menzies added the service station to his Richfield distributing plant. Today, it is the location of Denny's restaurant and a Chevron station. In 1936, William Mount rented the newly completed building on the left and established his implement business with the Case dealership. Later, the building housed the L.A. Hearne feed store. (Courtesy Casey family.)

AERIAL VIEW OF BRIDGES, 1956. The old and new Highway 101 bridges crossing the Salinas River are shown here. The old metal frame bridge is the lower one and was used for southbound traffic. The modern concrete bridge opened in June 1955 and carried northbound vehicles. Later, a concrete bridge was built just south of the other concrete one and southbound traffic was moved to it.

KING CITY ROOFTOP SIGN. This aerial view shows the town name on an oil company building on the east side of King City. At center is First Street, and on the right is the curve onto Broadway. The sign could have come about when the Mesa Del Rey Airport was opened. The first rooftop sign was installed on the El Camino Garage in March 1926 by owner William J. Hamilton.

Aerial View of Downtown. This view of Highway 101 shows it going through town with the Salinas River in the distance. At bottom is the 300 block of Broadway. The bend is at the intersection of Broadway and San Lorenzo Avenue, which was the original western boundary of the city. In this c. 1950 photograph, the cemetery is behind a row of eucalyptus trees.

Removal of El Topo Hotel Sign. In 1968, Highway 101 moved to the southwest edge of King City when the controversial K-line bypass was completed. Broadway then became much less used, with little need for through-traffic businesses such as gas stations, restaurants, or hotels. This photograph shows the removal of the El Topo Hotel sign at 333 Broadway depicting the change of use. (Courtesy Paul Layous of Tavernetti, Layous & Clark.)

Nine

Mesa Del Rey's Army Pilots

King City's airport played an important role in World War II in testing and training approximately 10,000 young men to be pilots.

It started when the Palo Alto Airport sought a site for one of their Army Air Force Contract Flying Schools. They discovered that the skies over the southern Salinas Valley offered conditions clear enough to train and fly nearly every day of the year.

In December 1940, the City of King subleased the airport land from Spreckels Sugar Company with an option to buy 249 acres around it. Local city leaders advocated for bringing the flight school to King City's airport, where the land adjacent to it became headquarters. By March 1941, the first cadets arrived. On October 14, 1941, an election was held for a bond measure to purchase the property.

As recruits arrived in King City there was a need for housing. With housing scarce, families were asked to rent rooms, and new apartments were built. There was a need for recreation, and the United Service Organizations (USO) quickly built a club on the 100 block of North Vanderhurst Avenue. Local young women in town often connected with these soldiers and some ended up marrying. Couples including Jeanette Rianda and Barney Clegg, Patricia Hayes and Marion Pope, and Helen Eade and Fred Carter stayed and raised families.

The flight school was the first stop for would-be pilots. If they were successful here for their nine-week stay, they moved on to more serious training and eventually did tours of duty as pilots in Europe and the Pacific. Records of those classes of trainees are preserved in yearbooks like those from high schools.

As quickly as the war ended, the troops left, and a new use for the airport property was needed. In 1957, Basic American Foods stepped in to buy a portion of the property from the city and convert it into an onion and garlic dehydrating plant. At one time, this business employed more than 700 year-round and seasonal workers.

The buildings have long acted as a beacon on the eastside bluff off Metz Road. This property went on to evolve into Conagra Foods and was later purchased in 2008 by the Rava family. Rava Ranches is among the largest farming operations in the valley.

Few towns the size of King City are fortunate to have an airport facility such as ours.

A display of the old airport sits in the current airport office, a gift from former pilot Mickey Muzinich. It is an ongoing reminder of the significance of Mesa Del Rey Airport.

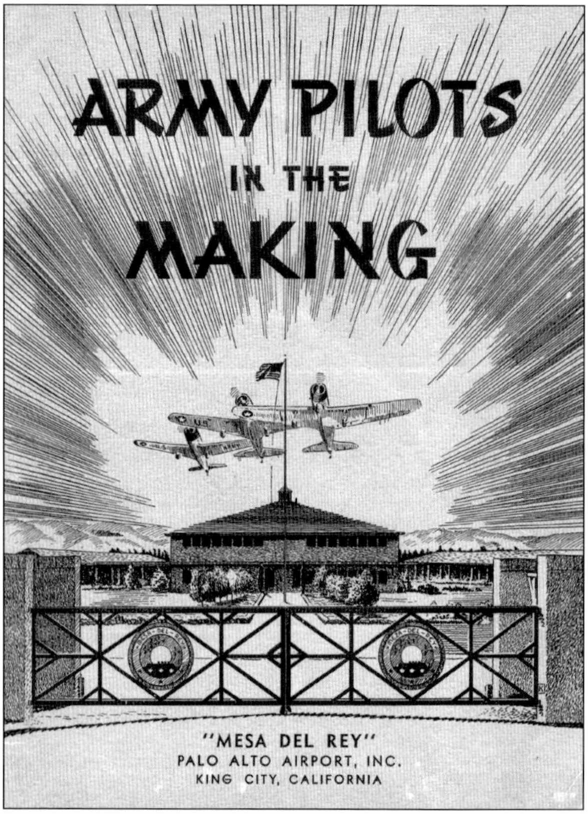

ARMY PILOTS IN THE MAKING. This promotional brochure contained information about the primary flight school at King City's Mesa Del Rey Airport. The cover shows a drawing of the main administration office constructed on the bench of land to the northeast of King City. Also constructed were barracks, a hospital, and airplane hangars. It was a private enterprise that contracted with the Palo Alto Airport.

ROWS OF AIRPLANES. This scene shows planes lined up during World War II at the Mesa del Rey Airport adjacent to Bitterwater Road. Note the line of eucalyptus trees commonly planted in the Salinas Valley to reduce wind. The website "Pilot's Log at Mesa Del Rey" tells some of the stories of events in King City during World War II. High school English teacher Anita Leonard worked with students to construct the site.

Revetments. At the top of this aerial view of Mesa Del Rey Airport during World War II are protective dirt berms known as revetments. They were used as protective shelter for airplanes in case of enemy air attacks. This view is on the east side of the airport with runways at center right. The buildings at bottom left in the pattern of a chevron contained barracks.

Mesa Del Rey Hangars. Structures like these went up quickly during World War II to prepare the Mesa del Rey airport property for the influx of soldiers. This photograph was part of a scrapbook belonging to trainer Fred Carter, who came to King City from Raleigh, North Carolina, to teach pilots. He married Helen Eade and stayed most of his life on their Long Valley ranch. (Courtesy Lise Carter for the Fred and Helen Carter family.)

USO Building. The structure at 121 North Vanderhurst was developed in 1941 for the Army Air Corps pilots who were training at Mesa Del Rey Airport. After the war, it became an American Legion hall. Bert Kerns, a local businessman, bought it and converted it for use as a roller skating rink and apartments. In 1977, Kerns sold it and it evolved into Buttgereit, Pettitt and Davis Agency and later the Robert Cullen Farmers Insurance Agency.

USO Building Dedication. Community members turned out to dedicate this building in December 1941 for recreation for those serving in the armed forces during World War II. The program included the King City High School band directed by Lowell Spencer, a welcome from Mayor Romilda Dedini, military officers from Hunter Liggett Military Reservation, and Palo Alto Airport Training School officials. It concluded with USO representative Charles Hogan turning the club over to the Salvation Army.

PAYDAY AT BANK OF AMERICA. Bank of America cashiers are busy rendering service to the aviation cadets from the Mesa Del Rey flight school. No doubt it is payday, as more than 30 are crowded into the bank's lobby at the southwest corner of Third Street and Broadway. In the *King's Log* yearbook for the class of 44-C, they are referred to as "Kings' Men."

DINING ROOM, SUK'S TAVERN. Here is a glimpse of the interior of this restaurant at 322 Broadway around 1942. A note on the back of the postcard mentions "where Peter and I had our last supper." This was likely a cadet from the flight school and his girlfriend during World War II. Her note continued, "Our attachment was suffering unattachment—it was like freedom and that was good!"

Airplane Spotting. The glass enclosure atop the El Camino Hotel is where volunteer spotters watched for enemy airplanes flying over during World War II. Workers had a book with illustrations of different planes, and binoculars. They logged the type of plane, the direction of the flights, and the time. Among those remembering the lookout were Josephine Koester, Marie Goodwin, and Benita Gray, granddaughter of "Ma" and "Pa" Hables.

Mesa Del Rey Flying Service. After the war, Mesa del Rey Flying Service focused on agriculture flights and crop dusting for pesticide application in the Salinas Valley. In recent years, Mesa Del Rey Airport has been home to the Tutima Academy of Aviation Safety, teaching the fundamentals of aerobatic flight.

Ten

EVERYDAY LIFE

People of all races and occupations have shaped King City.

A Frenchman named Charles King was the town's founder, but his birth name was LeRoi, a French word for "king."

One of Charles King's workers was a Chinese man named Mah Lon Sing. Sing became one of the early merchants in King City.

Spaniards and natives from Mexico were among the first to come on the Portola trek and with Juan Bautista De Anza. These early settlers had names such as Soberanes, Alviso, Castro, Garcia, Sanchez, and Soto. In the early 1900s, a Spaniard named Robert R. Diaz owned the Blackhawk Stables and served on the first city board of trustees.

Swiss and Italians dominated the dairy industry including families named Lombardi, Pozzi, Solari, and Tavernetti.

Coming from the Basque country were sheepherders with last names such as Echenique, Echeberria, and Errea.

German names one might recognize are Koester, Falkenberg, Bischoff, and Grab.

Portuguese names one might recognize are Pacheco, Souza, Silveira, and Domingos.

A large contingency of people from Oklahoma and Arkansas landed in King City during the Depression, with names such as Blackmon, Davis, and Bailey.

The best sources of early family histories are "mug books." In 1893, Barrows and Ingersoll published *Memorial Biographies*. In 1903, Prof. J.M. Guinn's mug books featured names such as J.M. Barbree, John N. Besse, John H. Bray, John Breen, Edward W. Kirk, and B.F. Norris. In 1925, S.J. Clarke Publishing Co. mentions W.F. Beasley, J.P. Wasson, William H.H. Metz, and Alvin Eade.

In the 1940s, the *King City Rustler* interviewed old-timers. Those articles were compiled into a book called *Salt of the Earth*. Businessman Jim Pettitt expanded on that by asking local families to write out their histories. He created a document called *Salt and Pepper of the Earth*.

At one time, the King City Chamber of Commerce used its Old Timers' Day annual gathering to compile information about local families.

In the 1970s, the San Antonio Valley Historical Association (SAVHA) took seriously the need to preserve old-timers' stories. Members interviewed and recorded on cassette tape the stories of more than 50 area residents. Now converted to CDs (and someday soon to electronic format), these stories can be heard at the Monterey County Free Library in King City, the archives of SAVHA, and the Nacitone Museum in Lockwood.

The space in this book is limited, but the following is a sample of some of the interesting people and places that have left an imprint on the town.

The Brunetti Family. Seated in front of their French Hotel on the 100 block of Broadway is Augustus Brunetti (right) and his wife, Maria Sanchez. At left is their son Julius, who was a merchant in King City for many years. Augustus and Maria first moved to the Hernandez Valley, east of King City in San Benito County, then moved to King City and built their hotel around 1887. (Courtesy Monterey County Historical Society.)

Spreckels House, 422 Lynn Street. Spreckels Sugar Company was known for this style of house supplied to workers on its ranches and in town. It is distinct because it is one of the only houses on Lynn Street that sits back from the curb. Note the rectangular window in the upper left-hand corner of this Harlan photograph from around 1908. These windows were typical of in-town houses; ranch houses had a stylized sugar beet ornament.

David "Paul" Talbott. This man had the distinction of serving as Monterey County supervisor for District No. 3 for 32 years from 1907 to 1939, the longest in county history. He was known as the "good roads" supporter. Supervisor Talbott is shown here with his wife, Isabelle Ellis. They lived on the Dunphy Ranch north of King City, next to the river where he raised grain, beans, and cattle. (Courtesy of Monterey County Historical Society.)

Hotel Reich Gathering. Residents of King City pose in this October 1914 photograph honoring the visit of Gov. Hiram Johnson, seated in the upholstered chair. The women from left to right are Sarah Vivian, wife of *Rustler* publisher Fred Vivian; Ida McReynolds, wife of city clerk A.B. McReynolds; and Mary Norris, wife of B.F. Norris. Also pictured are Daniel Harrington, Mayor Al Carlson, W.J. Hamilton, and Julius Brunetti. Hotel Reich was on the southwest corner of First Street and Broadway.

THE ULREY FAMILY. Lyman B. Ulrey, a pioneer businessman, is seen here with his daughter Alta seated in his 1910 EMF, one of the earliest cars in town. Standing in front of their residence at the northwest corner of Lynn Street and Vanderhurst Avenue are (from left to right) his wife, Ida, with daughters Vinnie and Esther. Ulrey was associated with Cooper Ulrey Company, King City Water Light and Power Company, and King City Garage. (Courtesy Butch LaVelle.)

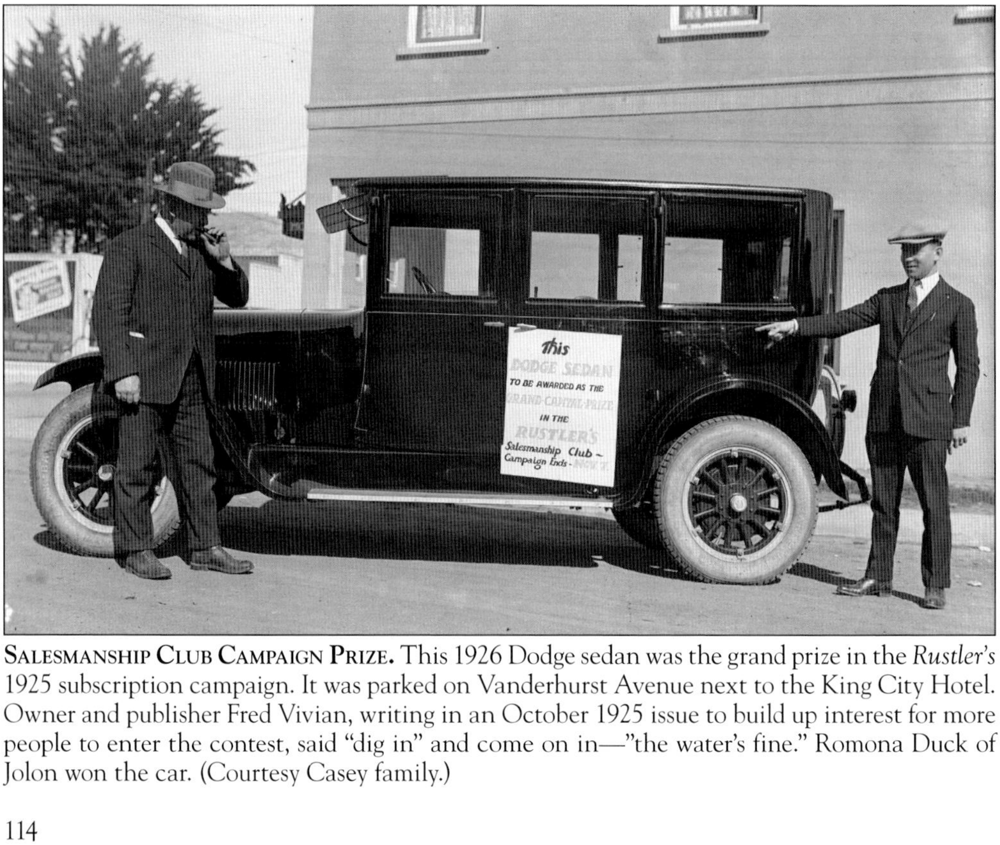

SALESMANSHIP CLUB CAMPAIGN PRIZE. This 1926 Dodge sedan was the grand prize in the *Rustler's* 1925 subscription campaign. It was parked on Vanderhurst Avenue next to the King City Hotel. Owner and publisher Fred Vivian, writing in an October 1925 issue to build up interest for more people to enter the contest, said "dig in" and come on in—"the water's fine." Romona Duck of Jolon won the car. (Courtesy Casey family.)

TRAFFIC OFFICER CLARENCE DODGE. This well-known 26-year-old traffic officer is shown sitting in front of L.E. "Doc" Cheney's Super-Service Garage at 401 Broadway near the corner of Vanderhurst Avenue. In October 1926, in pursuit of a speeder, he collided with a car at the Russ Avenue and Broadway intersection and died a short time later, becoming King City's first and only officer to die on duty. Officer Dodge was a native of Lockwood.

POLICE OFFICERS ON MOTORCYCLES. The King City Police Station was built in 1940 adjoining the old city jail at 114 North Third Street between Tynan Lumber Co. and Monterey County Bank. The building was constructed by contractor F.A. "Austin" Hayes. It included a room for Deputy Sheriff E.N. Harris. The fire department bell seen in the tower now hangs at the current fire department at 422 Bassett Street. (Courtesy Casey family.)

Cartoonist Eldon Dedini. Referred to as a "hometown boy who made good," Dedini was raised on a ranch outside King City and went on to be a nationally recognized cartoonist for magazines such as *Esquire* and *The New Yorker.* A King City High School graduate in 1939, he exhibited artistic talent at an early age, and while attending Salinas Valley Junior College in 1940–1941, his political drawings appeared frequently in the *Salinas Californian.* (Courtesy Monterey County Historical Society.)

King City Gardens Florist. Owner John Doyle sits in the panel truck used for flower deliveries. This scene is at 226 Ellis Street with his niece Roberta Thompson. John and his wife, Marie, entered the business in 1944 when it was located in the Reel Joy Theatre at 219 Broadway. On April 30, 1949, they held the grand opening of their 221 Bassett Street location. (Courtesy William T. "Bill" Doyle.)

Shirley and Al Francioli. Shirley Martella fell in love with and married Swiss dairyman Al Francioli before she graduated from King City High School in 1945. They lived in the country north of town until 1959, when they moved to 509 Bassett Street. For 34 of those years, from 1971 to 2005, Al and Shirley were beloved daycare providers for hundreds of children whose working parents relied on them. (Courtesy Shirley Francioli.)

The Daniels Family. This scene at 504 Lynn Street shows Pearl and Clarence Daniels in August 1942 with sons Danny (left) and Marvin. At the time, the Danielses lived in the home that was built for her grandparents Henry and Susanna "Ella" Rist, Lonoak-area ranchers. For 65 years, starting in 1937, Pearl kept a diary that was recently transcribed by Marvin. It provides a rare glimpse into everyday life in the King City area. (Courtesy Rist/Daniels family.)

ROTARY CLUB, 1937. In 1925, Rotary International decided to support a local club. Its first meeting was January 20, 1926, in El Camino Real Hotel with W.E. Goodspeed as the first president. On January 27, they held their charter banquet at the high school auditorium. According to the *Rustler*, "Hotel El Camino Real handled the culinary end." (Courtesy King City Rotary.)

DIME COLLECTION. Patrons at the Top Hat Bar used dimes to spell their names or businesses. Founded by Pres. Franklin Roosevelt in 1938, the March of Dimes asked children to donate dimes to find a cure for polio. Originally, containers with slotted tops were used, but that evolved into more creative displays. This is an example of the generosity for which this community is known.

King City Steam Laundry. This 1941 photograph shows the building at the southeast corner of Russ Avenue and Broadway with employees gathering in front. Work started on the concrete building in 1917, and it was opened on November 24, 1919, by S.S. Hill and Loren V. McKinsey. In March 1938, the building on the east was completed, and King City Cleaners moved in. (Courtesy Monterey County Agricultural and Rural Life Museum.)

Stampede Goodwill Air Tour. Cecilia LaVelle (left), a hostess for King City's 1940 Stampede, traded the horse for an airplane in this promotional tour. On the right is pilot Mary Hables, also of King City. This image shows them as they arrived at the Salinas airport. Romilda Dedini, mayor of King City, sponsored the two-day tour to cities of the Central Coast. (Courtesy Dan and Celie Hayes family.)

Dr. Dorus Brumwell and Wife. He was not King City's first doctor, but he was perhaps the town's most well-known physician in the first half of the 20th century. After his first wife, Emma Greve of Priest Valley, passed away in 1906, he married his housekeeper, Emy Petersen, in the summer of 1910. Dr. Brumwell would make house calls in rural areas on his Indian motorcycle or on foot if needed. He died in 1951 at age 90. (Courtesy Aileen Jones Brazelton.)

Andrus Connections. The Andrus family played an important role in providing medical services to King City. Dr. L.M. Andrus is shown in the portrait behind Alice Gaunt, a longtime nurse practitioner. He opened his practice in 1930, followed by opening a community hospital at 630 Broadway. In 1941, he opened a newer facility at 624 Broadway with 22 beds. His son L.H. Andrus (right) joined his father in the medical practice. (Courtesy Mee Memorial Healthcare System.)

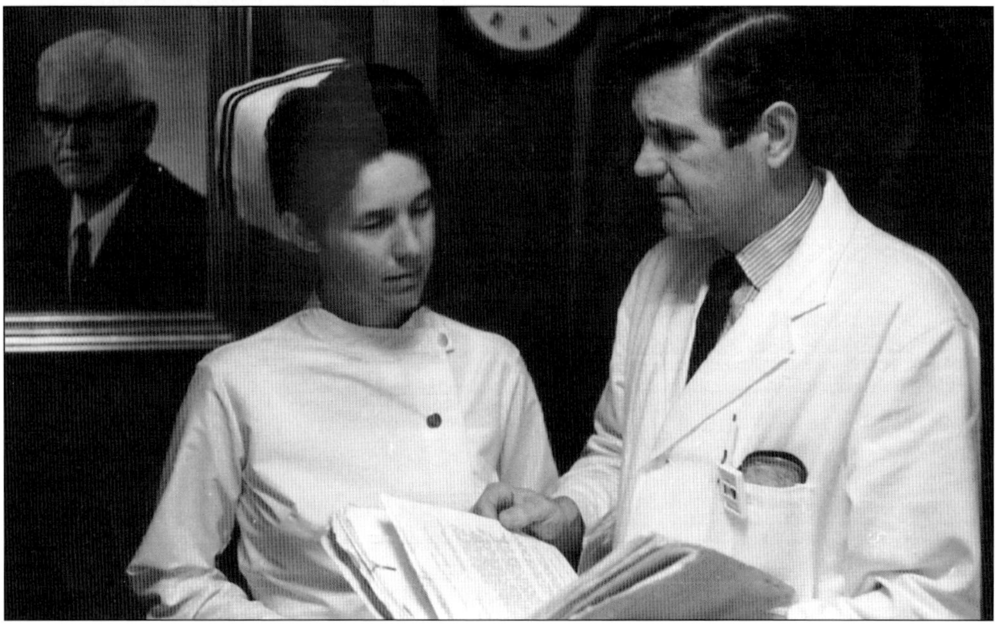

ASSEMBLYMAN PATTERSON. Ellis E. Patterson came to Monterey County to be the principal of Lockwood High School and San Antonio Grammar School. As a resident of King City, he ran for Assembly in 1932 and won. When running for re-election in 1936, he was defeated in the primary but staged a rare, successful write-in campaign. This notoriety helped propel him to be elected California's lieutenant governor in 1938. The story of his life is found in the book *Reflections of a California Liberal.* (Courtesy Patterson family.)

GOLF COURSE CLUBHOUSE. Since 1953, when it was built, the King City Golf Course has been an attraction for town residents and visitors. It is at the south end of Vanderhurst Avenue, where many of the fairways run parallel to and overlook San Lorenzo Creek. Operated by a private concessionaire, it hosts many tournaments. A brochure calls it "a diamond in the rough nine-hole public golf course." (Courtesy King City Golf Course.)

Salinas Valley Fair Carnival. Young and old alike have come to enjoy the rides and games at the annual fair on the 600 block of Division Street. The first Salinas Valley Fair was held in 1941 using part of the then King City Stampede grounds a block to the east on the 500 block of Division Street. The present site of the fair was purchased in 1946 from the Frank Tunzi family. (Courtesy Salinas Valley Fair.)

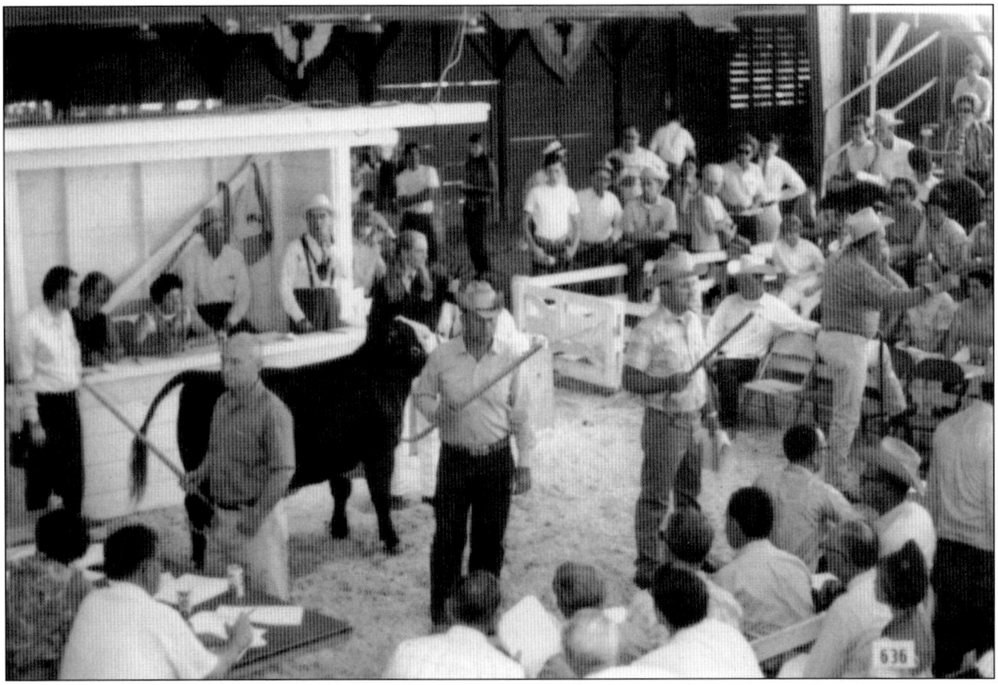

Salinas Valley Fair Livestock Auction. Action in the livestock arena is a key event at every Salinas Valley Fair. Recognizable here are Sam Avila Jr. (center) and Francis "Bosco" Echenique (right), who were staunch supporters of the fair. Though the Salinas Valley Fair started in 1941, it was not the first fair in King City. In 1924 and 1925, there were smaller fairs that used the Blackhawk Stables building on Broadway. (Courtesy Salinas Valley Fair.)

Pauline Hall Award. A descendent of early California families, Pauline Houx Hall brought dance into the lives of hundreds of young girls in her 36 years as the owner of the Pauline Hall Dance Studio and the producer of the annual Nutcracker Ballet starting in 1959. She is shown here with Salinas Valley Fair officials Ken Funk (left) and Maynard Frudden, acknowledging her support of the annual fair talent show. (Courtesy Salinas Valley Fair.)

Harry Casey with Tid. The daughter and grandson of *Rustler* founder Fred Vivian are shown here at 116 South Third Street, the newspaper's home for 40 years starting in 1937. Beatrice "Tid" Vivian Casey wrote articles for the paper and became a noted author. Harry F. Casey would stop by after school and help with setting type, running a press, or folding and wrapping newspapers. He became the publisher in 1952. (Courtesy Casey family.)

NEON KING CITY SIGN. This was the welcome sign for southbound traffic in the 1950s and 1960s. With the high school campus in the background, it was on the opposite side of Broadway from the cemetery. The Optimist club sign shows it met at Keefer's at 322 Broadway. The American Legion Post No. 67 also has its logo displayed. At the time, the chamber of commerce was at 509 Broadway.

Bibliography

Anderson, Burton. *From Valley to Sea: 25 Years with the Coastal Grower.* Salinas, CA: Monterey County Historical Society, 2015.

California Digital Newspaper Collection. cdnc.ucr.edu.

Clark, Donald Thomas. *Monterey County Place Names, A Geographical Dictionary.* Carmel Valley, CA: Kestrel Press, 1991.

Clovis, Margaret E. *Salinas Valley.* Charleston, SC: Arcadia Publishing, 2005.

Fisher, Anne B. *The Salinas Upside Down River.* New York, NY: Farrar & Rinehart, 1945.

Henry, G.P. "Early Settlement of the Town Described at Luncheon of Rotary Club." *King City Herald.* March 25, 1927.

King City, California: The First Hundred Years, 1886–1986. King City, CA: King City Centennial Committee, 1986. Reprinted 1999.

Kohrs, Donald G. *John Steinbeck's Roots.* 2021. https://seaside.stanford.edu/steinbeck.

Salt of the Earth. King City, CA: the *Rustler-Herald*, 1951.

Shepherd, Marvin. "Charles H. King: From Trinidad Educator to Oakland Millionaire." *Humboldt Historian.* Winter 2014.

Steinbeck, John. *East of Eden.* New York, NY: the Viking Press, 1952.

"The King Ranch Purchased by the Spreckels Sugar Company." *San Francisco Call.* December 9, 1897: 1

INDEX

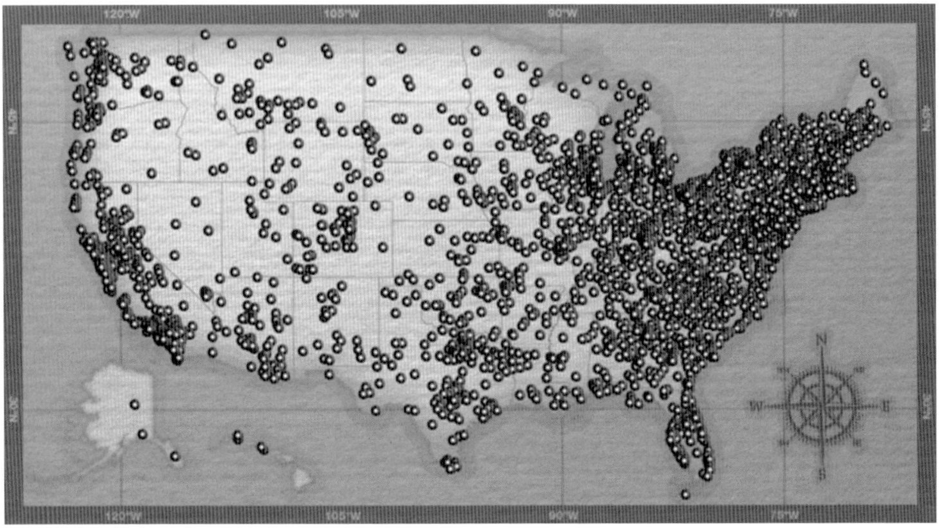